T0324126

THE KOREPIN FESTSCHRIFT

From Statistical Mechanics to Quantum Information Science

A Collection of Articles Written in Honor of the 60th Birthday of Vladimir Korepin

THE KOREPIN FESTSCHRIFT

From Statistical Mechanics to Quantum Information Science

A Collection of Articles Written in Honor of the 60th Birthday of Vladimir Korepin

Editors

Leong Chuan Kwek
(National University of Singapore & Nanyang Technological University, Singapore)

Simone Severini *(University College London, UK)*

Haibin Su *(Nanyang Technological University, Singapore)*

 **World Scientific**

NEW JERSEY · LONDON · SINGAPORE · BEIJING · SHANGHAI · HONG KONG · TAIPEI · CHENNAI

Published by

World Scientific Publishing Co. Pte. Ltd.

5 Toh Tuck Link, Singapore 596224

USA office: 27 Warren Street, Suite 401-402, Hackensack, NJ 07601

UK office: 57 Shelton Street, Covent Garden, London WC2H 9HE

British Library Cataloguing-in-Publication Data
A catalogue record for this book is available from the British Library.

THE KOREPIN FESTSCHRIFT: FROM STATISTICAL MECHANICS TO QUANTUM INFORMATION SCIENCE
A Collection of Articles Written in Honor of the 60th Birthday of Vladimir Korepin

Copyright © 2013 by World Scientific Publishing Co. Pte. Ltd.

All rights reserved. This book, or parts thereof, may not be reproduced in any form or by any means, electronic or mechanical, including photocopying, recording or any information storage and retrieval system now known or to be invented, without written permission from the Publisher.

For photocopying of material in this volume, please pay a copying fee through the Copyright Clearance Center, Inc., 222 Rosewood Drive, Danvers, MA 01923, USA. In this case permission to photocopy is not required from the publisher.

ISBN 978-981-4460-31-6

Printed in Singapore

PREFACE

On 25–28 May 2011, the Institute of Advanced Studies at the Nanyang Technological University organized the fifth Asia-Pacific Workshop on Quantum Information Science (APWQIS) in conjunction with a Festchriff in honor of Vladimir Korepin's sixtieth birthday. Although Vladimir has made immense contribution to condensed matter physics and integrable models, he has recently forayed into the nascent field of quantum information science.

Vladimir Korepin completed his undergraduate study at Saint Petersburg State University. He was subsequently employed by the Mathematical Institute of Academy of Sciences of Russia. He completed his graduate and postdoctoral studies at this Institute in 1977 and he continued to work there till 1989. His scientific adviser was Ludwig Faddeev. In 1985, he received a doctor of sciences degree in mathematical physics from the Council of Ministers of the Russian Federation.

Vladimir has written several scholarly books. In particular, he is well known for his book on one-dimensional Hubbard model. He wrote the book with Fabian Essler, Holger Frahm, Frank Goehmann and Andreas Kluemper. Together with Nikolay Bogolyubov and A. G. Izergin, they have also published a book under the Cambridge University Press on quantum inverse scattering method. Most of Vladimir's excellent work on the massive Thirring model can also be found in this text.

In mathematical physics, Vladimir has introduced domain wall boundary conditions for the six vertex model.[1] The application of such boundary conditions has appeared in several diverse fields of mathematics such as algebraic combinatorics, alternating sign matrices, domino tiling, Young diagrams and plane partitions. Vladimir has also initiated the study of differential equations for quantum correlation functions and showed that these functions can lead to the discovery of a special class of Fredholm integral operators. They now have multiple applications not only to quantum exactly solvable models, but also to random matrices and algebraic combinatorics.

In quantum information theory, Vladimir has produced several important results regarding the analytical evaluation of the entanglement entropy of different dynamical models, such as interacting spins, Bose gases, and the Hubbard model.[2] He considered models with a unique ground states, so that the entropy of the whole ground state is zero. The ground state is partitioned into two spatially separated parts: the block and the environment. He calculated the entropy of the block as a function of its size and other physical parameters. In a series of articles,.[3–9] He was also the first to compute an analytic formula for the entanglement entropy of the XX (isotropic) and XY Heisenberg models based on Toeplitz Determinants and Fisher–Hartwig formula. He has also made important contributions to Valence–Bond–Solid states (which is the ground state of the Affleck–Kennedy–Lieb–Tasaki model of interacting spins) and quantum search algorithms with Lov Grover.[10,11]

In this Festchriff, we have assembled a medley of interesting articles from some of his friends, well-wishers and collaborators. It would be hard to describe all of Vladimir's contribution in a small volume. Last but not least, we wish him a happy sixtieth birthday and we hope that he will continue to contribute to science and the community for many more years.

References

1. V. E. Korepin, *Commun. Math. Phys.* **86**, 191 (1982).
2. V. E. Korepin, *Phys. Rev. Lett.* **92**, 096402 (2004).
3. B.-Q. Jin and V. E. Korepin, *J. Stat. Phys.* **116**, 79 (2004).
4. A. R. Its, B.-Q. Jin and V. E. Korepin, *J. Phys. A: Math. Gen.* **38**, 2975 (2005).
5. A. R. Its, B.-Q. Jin and V. E. Korepin, Entropy of *XY* Spin Chain and Block Toeplitz Determinants, *Fields Institute Communications, Universality and Renormalization,* eds. I. Bender and D. Kreimer, **50**, 151 (2007).
6. F. Franchini *et al.*, *J. Phys. A: Math. Theor.* **40**, 8467 (2007).
7. F. Franchini, A. R. Its and V. E. Korepin, *J. Phys. A: Math. Theor.* **41**, 025302 (2008).
8. H. Fan, V. E. Korepin and V. Roychowdhury, *Phys. Rev. Lett.* **93**, 227203 (2004).
9. V. E. Korepin and Y. Xu, Entanglement in Valence–Bond–Solid States, arXiv:quant-ph/0908.2345.
10. V. E. Korepin and L. K. Grover, Simple Algorithm for Partial Quantum Search, arXiv:quant-ph/0504157.
11. V. E. Korepin and B. C. Vallilo, Group Theoretical Formulation of Quantum Partial Search Algorithm, arXiv:quant-ph/0609205.

Editors

Leong Chuan Kwek
National University of Singapore
Singapore

Simone Severini
University College London
UK

Haibin Su
Nanyang Technological University
Singapore

CONTENTS

Preface   v

Organizing Committee   ix

List of Speakers   xi

Photos   xiii

Chapter 1: Cancellation of Ultra-Violet Infinities in One Loop Gravity   1
    V. E. Korepin

Chapter 2: Quantum Discord in a Spin System with Symmetry Breaking   11
    B. Tomasello, D. Rossini, A. Hamma and L. Amico

Chapter 3: Entanglement from the Dynamics of an Ideal Bose Gas in a
Lattice   29
    S. Bose

Chapter 4: Aspects of the Riemannian Geometry of Quantum
Computation   37
    H. E. Brandt

Chapter 5: Quantum Mechanics and the Role of Time: Are Quantum
Systems Markovian?   61
    T. Durt

Chapter 6: Explicit Formula of the Separability Criterion for Continuous
Variables Systems   77
    K. Fujikawa

Chapter 7: Yang–Baxter Equations in Quantum Information   85
    M.-L. Ge and K. Xue

Chapter 8: Nondistillable Entanglement Guarantees Distillable
Entanglement   105
    L. Chen and M. Hayashi

Chapter 9: Reduced Density Matrix and Entanglement Entropy of
Permutationally Invariant Quantum Many-Body Systems   119

V. Popkov and M. Salerno

Chapter 10: Solitons Experience for Black Hole Production in
Ultrarelativistic Particle Collisions   141
  I. Ya. Aref'eva

Chapter 11: Sine–Gordon Theory in the Repulsive Regime,
Thermodynamic Bethe Ansatz and Minimal Models   161
  H. Itoyama

Chapter 12: On Some Algebraic and Combinatorial Properties of Dunkl
Elements   171
  A. N. Kirillov

Chapter 13: Finite Projective Spaces, Geometric Spreads of Lines and
Multi-Qubits   199
  M. Saniga

Chapter 14: Monogamy of Entanglement, N-Representability Problems
and Ground States   203
  T.-C. Wei

Scientific Programs   211

List of Participants   215

ORGANIZING COMMITTEE

International Advisory Committee

Rodney BAXTER   (Australian National University)
Howard CARMICHAEL   (University of Auckland)
Ludwig FADDEEV   (Steklov Institute of Mathematics)
Kazuo FUJIKAWA   (Nihon University)
Molin GE   (Nankai University)
Jaewan KIM   (Korea Institute for Advanced Study)
Choy Heng LAI   (National University of Singapore)
Franco NORI   (RIKEN Advanced Science Institute)
Kok Khoo PHUA   (Nanyang Technological University)
Jason TWAMLEY   (Macquarie University)
Chen Ning YANG   (Nobel Laureate in Physics, 1957)

Local Organizing Committee

Kok Khoo PHUA   (Nanyang Technological University)
Choo Hiap OH   (National University of Singapore)
Jiangbin GONG   (National University of Singapore)
Choy Sin HEW   (Nanyang Technological University)
Leong Chuan KWEK   (Nanyang Technological University)
Hwee Boon LOW   (Nanyang Technological University)
Simone SEVERINI   (University College London)
Haibin SU   (Nanyang Technological University)
Chengjie ZHANG   (National University of Singapore)

LIST OF SPEAKERS

Luigi AMICO,   Uni-Catania, Italy

Rodney James BAXTER,   Australian National University, Australia

Sougato BOSE,   University College London, UK

Howard E. BRANDT,   U.S. Army Research Laboratory, USA

Howard CARMICHAEL,   University of Auckland, New Zealand

Darrick E. CHANG,   California Institute of Technology, USA

CHEN Yan,   Fudan University, China

CHEN Zhanghai,   Fudan University, China

DU Jiangfeng,   University of Science and Technology of China, China

Thomas DURT,   Institut Fresnel, Ecole Centrale de Marseille, France

Kazuo FUJIKAWA,   Nihon University, Japan

GE Molin,   Nankai University, China

GOAN Hsi-Sheng,   National Taiwan University, Taiwan

GUAN Xiwen,   Australian National University, Australia

HAN Zhengfu,   University of Science and Technology of China, China

Masahito Hayashi,   Tohoku University, Japan

HU Bei-Lok,   University of Maryland, USA

JIN Baiqi,   Wenzhou University, China

Hosho KATSURA,   Gakushuin Univeristy, Tokyo, Japan

Jaewan KIM,   Korea Institute for Advanced Study, Korea

Vladimir KOREPIN,   State University of New York, USA

LAM Ping Koy,   The Australian National University, Australia

LI Fuli,   Xi'an Jiaotong University, China

LIN Hai-Qing,   The Chinese University of Hong Kong, Hong Kong SAR (China)

LIU Wuming,   The Chinese Academy of Sciences, China

LIU Renbao,   The Chinese University of Hong Kong, Hong Kong

LUO Shunlong,   Academy of Mathematics and Systems Science, Chinese Academy
of Sciences, China

Franco NORI,   RIKEN Advanced Science Institute and University of Michigan,
Japan

Vladislav Popkov,   University of Salerno, Italy

Barry SANDERS,   University of Calgary, Canada

Simone SEVERINI,   University College London, UK

Alexei TSVELIK,   Brookhaven National Laboratory, USA

Thibault VOGT,   Peking University, China

Igor VOLOVICH,   Steklov Mathematical Institute (Russian Academy of Sciences),
Russia

WANG Zidan,   The University of Hong Kong, Hong Kong
WANG Xiangbin,   Tsinghua University, China
Paul Wiegmann,   University of Chicago, USA
Howard WISEMAN,   Griffith University, Australia
YAO Wang,   The University of Hong Kong, Hong Kong
YOU Jianqiang,   Fudan University, China
YU Yang,   Nanjing University, China
YU Sixia,   CQT/National University Singapore, Singapore
ZHANG Weiping,   East China Normal University, China
ZHANG Wei-Min,   National Cheng Kung University, Taiwan
ZHANG Tiancai,   Shanxi University, China

**5th Asia-Pacific Workshop on
Quantum Information Science 2011
in conjunction with the Festschrift
in honour of Vladimir Korepin**

NUS
National University
of Singapore
Department of Physics

NANYANG
TECHNOLOGICAL
UNIVERSITY
Institute of Advanced Studies

Korepin was trying to reply to a question that somebody was asking during the talk.

Korepin giving his talk.

Korepin giving his talk.

Korepin giving his talk.

Korepin with his birthday cake ... giving his birthday speech after the Russian Birthday song.

Korepin blowing candles after his speech.

Igor Volovich giving his talk on quantum photosynthesis.

Vladislav Popkov giving his talk entanglement of permutational invariant quantum states.

Masahito Hayashi showing how weaker entanglement could guarantees stronger entanglement.

Kazuo Fujikawa giving his talk on separability criterion.

Rodney Baxter delivering his talk on Ising and chiral Potts models.

Tsvelik giving his talk on realization of zero energy Majorana modes in spin ladders.

Luo Shunlong talking about measurement induced nonlocality.

Xiangbin Wang on improving quantum entanglement through single-qubit operation.

Chapter 1

CANCELLATION OF ULTRA-VIOLET INFINITIES IN ONE LOOP GRAVITY

V. E. KOREPIN

C. N. Yang Institute for Theoretical Physics,
State University of New York at Stony Brook,
Stony Brook 11794-3840, New York

Quantization of gravity is a formidable challenge for modern theoretical physics. Ultraviolet divergencies are the problem. In 1974 the author used functional integral formalism for the quantization. The background formulation helped to solve the problem in one loop approximation. The author proved that the divergencies cancel on mass shell, because of Einstein's equations of motion.

Keywords: Gravity, quantization.

1. Introduction

This is a historical note. In 1974 I was an undergraduate student of L. D. Faddeev. I was working on quantum gravity [without matter] in one loop approximation. I discovered [simultaneously with G. t'Hooft M. Veltman] that on mass shell ultraviolet divergences cancel. The text below is a translation of the Diploma.

The diploma was in Russian, it was never published. Recently Faddeev found a reference to my result by an editor in Feynman's lectures on gravitation[1] (the reference is on page xxxvii of the book). The scan of original diploma can be found in the first line of the webpage.[2] The result was obtained simultaneously with G. t'Hooft M. Veltman.[3] The translation of the diploma follows.[a]

The diploma is devoted to quantization of gravity. First appropriate method for quantization of gravity was formulated by Dirac[7] in 1958 in a frame of Hamiltonian approach. Other methods [essentially equivalent to Dirac's] were suggested in Refs. 8–13. Hamiltonian approach is not covariant, this makes perturbation theory complicated.

An approach to covariant quantization of gravity was suggested in Refs. 14 and 15, but this method led to violation of unitarity.[16] In the same publication Feynman showed that to restore unitarity of a diagram in a form of a closed ring one has to

---

[a]Fictitious particles mentioned in the Diploma now called ghosts.

substract another diagram also in a form of a ring, which describes propagation of fictitious particle. Solution of the problem for any diagram was formulated in 1967 by Fadeev and Popov[20] and De-Witt[17-19] [using essentially different approaches]. This approach to quantization contains fictitious particle [ghosts] and their interaction with gravitons. Next question is renormalization. Formally the whole theory is not renormalizable. Here the pure gravity is studied in one loop approximation.

2. Covariant Quantization

We shall use functional integral for quantization of gravity. The background formulation for generating functional for scattering matrix will be used, not generating functional of Green functions. The integral will depend on asymptotic fields. We shall consider scattering matrix directly on mass shell, using classical equation of motion. Matrix element of scattering matrix with n external lines can be obtained by n multiple differentiation of our generating functional with respect to asymptotic fields. Asymptotic fields are classical fields. The generating functional of scattering matrix covariantly depends on asymptotic fields. We will see that some divergencies disappear on mass shell.

Let us denote by $g_{\mu\nu}^{\text{cl}}$ a classical solution of Einstein equations.[b] So $R = 0$ and $R_{\mu\nu} = 0$.

Let $iW(g_{\mu\nu}^{\text{cl}})$ denote a generating functional for connected Feynman diagrams with loops:

$$e^{iW} = e^{-iS_{\text{Gr}}(g_{\rho\omega}^{\text{cl}})} \int \exp i S_{\text{Gr}}(g_{\rho\omega}^{\text{cl}} + u_{\rho\omega}) \det\left(-g_{\gamma\delta}^{\text{cl}} - u_{\gamma\delta}\right)^{-5/2} d^{10}u_{\alpha\beta} . \tag{1}$$

Here

$$S_{\text{Gr}}(g_{\alpha\beta}) = \int \sqrt{g} R(g_{\alpha\beta}) d^4 x \tag{2}$$

is Einstein action for gravitational field. The measure $g^{-5/2} d^{10} g_{\mu\nu}$ assures unitarity. The formula (1) formally covariant under nonabelian group of coordinate transformation. The theory is similar to nonabelian gauge field. We shall use Faddeev–Popov approach to quantization, also we should choose a gauge to keep explicit covariance of W. Below we shall use notations:

$$g_{\mu\nu}^t = g_{\mu\nu}^{\text{cl}} + u_{\mu\nu}, \quad g_t^{\alpha\beta} g_{\beta\mu}^t = \delta_\mu^\alpha . \tag{3}$$

Also ∇^{cl} will be a covariant derivative with respect to $g_{\mu\nu}^{\text{cl}}$ (sometimes we shall write cl as a lower index) and ∇^t will be a covariant derivative with respect to $g_{\mu\nu}^t$ (sometimes we shall write t as a lower index).

Let us introduce an auxiliary condition:

$$\nabla_{\text{cl}}^\mu U_{\mu\nu} - \frac{1}{2} g_{\text{cl}}^{\alpha\beta} \nabla_\nu^{\text{cl}} U_{\alpha\beta} - C_\nu = 0 , \tag{4}$$

---

[b]In case of upper indices we shall write $g_{\text{cl}}^{\mu\nu}$.

here C_{nu} is an arbitrary function of coordinates. We shall also use:

$$D_\nu = \nabla^\mu_{\text{cl}} U_{\mu\nu} - \frac{1}{2} g^{\alpha\beta}_{\text{cl}} \nabla^{\text{cl}}_\nu U_{\alpha\beta} \,. \tag{5}$$

So in one loop quantization we get:

$$e^{iW} = \int \exp i S_{\text{Gr}}(g^t_{\mu\nu}) \det \hat{M} \prod_x \delta^4(D_\nu - C_\nu) g_t^{-5/2} d^{10} U_{\mu\nu} \,. \tag{6}$$

We used $S_{\text{Gr}}(g^{\text{cl}}_{\alpha\beta}) = 0$. The one loop operator is

$$\hat{M}_{\nu\omega} = \nabla^\mu_{\text{cl}} \nabla^t_\mu g^t_{\nu\omega} + \nabla^\mu_{\text{cl}} \nabla^t_\nu g^t_{\mu\omega} - g^{\alpha\beta}_{\text{cl}} \nabla^{\text{cl}}_\nu \nabla^t_\alpha g^t_{\beta\omega} \,. \tag{7}$$

It is checked in the original diploma that the right hand side of (6) transforms as scalar density under coordinate transformations. At this point we want to emphasize that $g^{\text{cl}}_{\mu\nu}$ is an arbitrary solution of Einstein equations. Let us get rid of the δ function following ideas of t'Hooft.[6] Note that the expression

$$N = \left(\prod_x \sqrt{g_{\text{cl}}} \right) \int dC_\nu \exp\left\{ -i \int d^4x \sqrt{g^{\text{cl}}} g^{\alpha\beta}_{\text{cl}} C_\alpha C_\beta \right\} \tag{8}$$

does not depend on $g^{\text{cl}}_{\mu\nu}$. The constant factor in the Eq. (6) does not matter, so we can multiply the right hand side by N.

$$e^{iW} = \left(\prod_x \sqrt{g_{\text{cl}}} \right) \int dC_\nu \exp\left\{ -i \int d^4x \sqrt{g^{\text{cl}}} g^{\alpha\beta}_{\text{cl}} C_\alpha C_\beta \right\}$$
$$\times \int \exp i S_{\text{Gr}}(g^t_{\mu\nu}) \det \hat{M} \prod_x \delta^4(D_\nu - C_\nu) g_t^{-5/2} d^{10} U_{\mu\nu} \,. \tag{9}$$

After integrating with respect to C_ν we obtain:

$$e^{iW} = \left(\prod_x \sqrt{g_{\text{cl}}} \right) \int \exp\left\{ i S_{\text{Gr}}(g^t_{\mu\nu}) - i \int d^4x \sqrt{g^{\text{cl}}} g^{\alpha\beta}_{\text{cl}} D_\alpha D_\beta \right\} (\det \hat{M}) g_t^{-5/2} d^{10} U_{\mu\nu} \,. \tag{10}$$

We can also represent $\det \hat{M}$ as an integral with respect to anti-commuting vector fields $\bar{\chi}^\alpha$

$$\det \hat{M} = \int d^4 \bar{\chi}^\alpha d^4 \chi^\beta \exp i \int d^4x \bar{\chi}^\alpha M_{\alpha\beta} \chi^\beta \,. \tag{11}$$

Finally we arrive to the following expression for generating functional for scattering matrix:

$$e^{iW} = \left(\prod_x \sqrt{g_{\text{cl}}} \right) \int \exp\left\{ i S_{\text{Gr}}(g^t_{\mu\nu}) - i \int d^4x \sqrt{g^{\text{cl}}} g^{\alpha\beta}_{\text{cl}} D_\alpha D_\beta \right\}$$
$$\times \exp\left(i \int d^4x \bar{\chi}^\alpha \hat{M}_{\alpha\beta} \chi^\beta \right) g_t^{-5/2} d^{10} U_{\mu\nu} d^4 \bar{\chi}^\alpha d^4 \chi^\beta \,. \tag{12}$$

We do not care about common factors. We can calculate the integral by stationary phase approximation. The stationary point is:

$$g_{\mu\nu} = g_{\mu\nu}^{\text{cl}}, \quad \bar{\chi}^\alpha = 0, \quad \chi^\beta = 0. \tag{13}$$

We shall leave in the exponent only terms quadratic in integration variables. In this approximation we get:

$$\bar{\chi}^\alpha \hat{M}_{\alpha\beta} \chi^\beta = \bar{\chi}^\alpha g_{\alpha\beta}^{\text{cl}} \nabla_{\text{cl}}^\mu \nabla_\mu^{\text{cl}} \chi^\beta. \tag{14}$$

The integral

$$e^{iW} = \left(\prod_x g_{\text{cl}}^{-1} \right) \int \exp \left\{ iU_{\mu\nu} \frac{\delta^2 S}{\delta g_{\mu\nu}^{\text{cl}} \delta g_{\alpha\beta}^{\text{cl}}} U_{\alpha\beta} - i \int d^{4\mu} x \sqrt{g^{\text{cl}}} g_{\text{cl}}^{\alpha\beta} D_\alpha D_\beta \right\}$$

$$\times \exp \left(i \int d^4 x \bar{\chi}^\alpha \nabla_{\text{cl}}^\mu \nabla_\mu^{\text{cl}} \chi^\alpha \right) d^{10} U_{\mu\nu} d^4 \bar{\chi}^\alpha d^4 \chi^\beta \tag{15}$$

can be taken. Now we have to calculate the quadratic form:

$$U_{\mu\nu} \frac{\delta^2 S}{\delta g_{\mu\nu}^{\text{cl}} \delta g_{\alpha\beta}^{\text{cl}}} U_{\alpha\beta} - \int d^{4\mu} x \sqrt{g^{\text{cl}}} g_{\text{cl}}^{\alpha\beta} D_\alpha D_\beta = \frac{1}{2} \int d^4 x U_{\mu\nu} F_{\alpha\beta}^{\hat{\mu}\nu} {}_{\text{cl}} g_{\text{cl}}^{\alpha\lambda} g_{\text{cl}}^{\beta\delta} U_{\lambda\delta}. \tag{16}$$

In the rest of diploma we shall use only $g_{\alpha\beta}^{\text{cl}}$, so we shall drop index cl. For $F_{\alpha\beta}^{\hat{\mu}\nu}{}_{\text{cl}}$ we get:

$$F_{\alpha\beta}^{\hat{\mu}\nu}{}_{\text{cl}} = \frac{1}{2} (\delta_\rho^\mu \delta_\lambda^\nu + \delta_\lambda^\mu \delta_\rho^\nu - g^{\mu\nu} g_{\rho\lambda})(\nabla^\theta \nabla_\theta \delta_\alpha^\rho \delta_\beta^\lambda - 2R_{\alpha\beta}^{\rho\lambda}). \tag{17}$$

It is convenient to denote:

$$\hat{F}_v = (\nabla^\theta \nabla_\theta)_{(v)}. \tag{18}$$

We can evaluate Gaussian integrals in the form (15)

$$e^{iW} = \left(\prod_x g_{\text{cl}}^{-1} \right) (\det^{-1/2} \hat{F}_{\text{Gr}}) \det \hat{F}_f. \tag{19}$$

This is the expression for generating functional of scattering matrix in one loop approximation.

3. Cancelation of Infinities in one Loop Approximation

To calculate determinants in formula (19) we shall use method of proper time

$$\ln \det F = -\text{Tr} \int_0^\infty \frac{ds}{s} (e^{i(F+i0)s} - e^{isI}). \tag{20}$$

We are going to differentiate the left hand side, so $\text{Tr} \int_0^\infty (ds/s) e^{isI}$ will not contribute and we shall not write it.

Let us write differential equation and initial data:

$$\frac{\partial e^{i\hat{F}s}}{\partial s} = i\hat{F} e^{i\hat{F}s}, \quad e^{i\hat{F}s}|_{s=0} = I. \tag{21}$$

Let us denote by $G(x, y|s)$ the kernel of the integral operator $\exp(i\hat{F}s)$. The main part of the operator $\hat{F}$ is d'Alembertian. So we can separate a singular factor characteristic for parabolic equation:

$$G(x, x'|s) = \frac{-1}{(4\pi s)^2} \exp\left(\frac{i\sigma(x, x')}{2s}\right) D^{1/2} A(x, x'|s). \tag{22}$$

Here $A(x, x'|s)$ is a smooth function which turns into 1 at $s = 0$. The $\sigma(x, x')$ is geodesic distance between points x and x'. It satisfies a differential equation:

$$g^{\mu\nu} \partial_\mu \sigma(x, x') \partial_\nu \sigma(x, x') = 2\sigma(x, x'). \tag{23}$$

In case of flat space $2\sigma(x, x') = (x - x')^2$. The third factor in (22) is density $D(x, x') = \det(-\sigma_{\mu\nu'}(x, x'))$. It satisfies a differential equation:

$$D^{-1}(\sigma^\mu D)._\mu = 4. \tag{24}$$

It is also convenient to introduce a scalar:

$$\Delta = \frac{D}{\sqrt{g}\sqrt{g'}}.$$

We use the following notation: $\partial_\nu \Phi = \partial\Phi/\partial x^\nu$ and $\partial_{\nu'} \Phi = \partial\Phi/\partial x'^\nu$. Later we shall use $\lim_{x \to x'} D = g(x)$. This follows from $\lim_{x \to x'} \partial_\mu \partial_{\nu'} \sigma(x, x') = g_{\mu\nu}(x)$. In order to calculate $\exp i\hat{F}s$ we need to introduce a function of parallel transport $g^\alpha_{\beta'}(x, x')$. It is a by-vector: index α is related to the point x and β' to x'. The function satisfy the equation: $\sigma^\tau g^\alpha_{\beta'.\tau} = 0$. Corresponding boundary condition is $g^\alpha_{\beta'}(x, x') \to \delta^\alpha_{\beta'}$ as $x \to x'$.[c] The function of parallel transport has the following properties:

$$g_{\mu\nu'} = g_{\nu'\mu}, \quad g^{\nu'}_\mu \sigma._{\nu'} = -\sigma._\mu, \quad g_{\mu\sigma'} g^{\sigma'}_\nu = g_{\mu\nu}, \quad \det(-g_{\mu\nu'}) = \sqrt{gg'}.$$

Now we are ready to study the formula (22). For fictitious particles we put

$$A(x, y|s)^\alpha_{\beta'} = g^\alpha_{\beta'}(x, y) f^f(x, y|s). \tag{25}$$

The function $f^f(x, y|s)$ is a by-scalar satisfying equation:

$$\frac{\partial f^f}{\partial s} + \frac{\sigma.\mu f^f._\mu}{s} = \frac{i}{4} g^\alpha_{\beta'} \Delta^{-1/2} \nabla^\mu \nabla_\mu (\Delta^{1/2} g^{\beta'}_\alpha f^f). \tag{26}$$

Consider Taylor series:

$$f^f(x, y|s) = \sum_{n=0}^{\infty} a^f_n(x, y)(is)^n, \quad a_0 = 1. \tag{27}$$

Coefficients satisfy equations:

$$\sigma^\mu a_{n.\mu} + n a_n = \frac{1}{4} \Delta^{-1/2} g^\alpha_{\beta'} (g^{\beta'}_\alpha \Delta^{1/2} a_{n-1})^\theta._\theta. \tag{28}$$

---

[c] In flat space $g^\alpha_{\beta'}(x, x') = \delta^\alpha_{\beta'}$.

Let us do similar calculations for gravitons:

$$A(x,y|s)^{\mu\nu}_{\alpha'\beta'} = \frac{1}{2}\left(g^{\mu}_{\lambda'}g^{\nu}_{\gamma'} + g^{\mu}_{\gamma'}g^{\nu}_{\lambda'} - \frac{1}{2}g^{\mu\nu}g_{\gamma'\lambda'}\right)f^{\gamma'\lambda'}_{\alpha'\beta'}(x,y|s). \tag{29}$$

Here f is a scalar at x and 4-tensor at point y, it is symmetric and traceless with respect to $\gamma'\lambda'$. It satisfies an equation:

$$\frac{\partial}{\partial s}f^{\mu'\nu'}_{\alpha'\beta'} + \frac{\sigma^{\omega}_{.\omega}}{s}(f_{.\omega})^{\mu'\nu'}_{\alpha'\beta'} = ig^{\mu'}_{\omega}g^{\nu'}_{\delta}\Delta^{-1/2}(\Delta^{1/2}g^{\omega}_{\lambda'}g^{\delta}_{\nu'}f^{\lambda'\nu'}_{\alpha'\beta'})^{\theta}_{.\theta}$$

$$- 2ig^{\mu'}_{\theta}g^{\nu'}_{\varsigma}R^{\theta\varsigma}_{\omega\delta}g^{\omega}_{\lambda'}g^{\delta}_{\gamma'}f^{\lambda'\gamma'}_{\alpha'\beta'}. \tag{30}$$

Consider Taylor series for this f

$$f^{\mu'\nu'}_{\alpha'\beta'} = \sum_{n=0}^{\infty}a^{\mu'\nu'}_{\alpha'\beta'}(is)^n. \tag{31}$$

Coefficients satisfy a recursion:

$$\sigma^{\omega}_{.\omega}(a_{n.\omega})^{\mu'\nu'}_{\alpha'\beta'} + n(a_n)^{\mu'\nu'}_{\alpha'\beta'} = g^{\mu'}_{\omega}g^{\nu'}_{\delta}\Delta^{-1/2}(\Delta^{1/2}g^{\omega}_{\lambda'}g^{\delta}_{\nu'}(a_{n-1})^{\lambda'\nu'}_{\alpha'\beta'})^{\theta}_{.\theta}$$

$$- 2g^{\mu'}_{\theta}g^{\nu'}_{\varsigma}R^{\theta\varsigma}_{\omega\delta}g^{\omega}_{\lambda'}g^{\delta}_{\gamma'}(a_{n-1})^{\lambda'\gamma'}_{\alpha'\beta'}. \tag{32}$$

Note that the coefficients $(a_n)^{\lambda'\gamma'}_{\alpha'\beta'}$ are symmetric and traceless with respect to upper indices $\lambda'\gamma'$. Also

$$(a_0)^{\lambda'\gamma'}_{\alpha'\beta'} = \frac{1}{2}\left(\delta^{\lambda'}_{\alpha'}\delta^{\gamma'}_{\beta'} + \delta^{\lambda'}_{\beta'}\delta^{\gamma'}_{\alpha'} - \frac{1}{2}g^{\lambda'\gamma'}g_{\alpha'\beta'}\right). \tag{33}$$

These calculations directly generalize the ones by De Witt [he used them for description of interaction of scalar particles with external gravity]. So we described the kernel $\exp(is\hat{F})$, see Eq. (22). We can use it to separate infinities in the formula (20). Ultraviolet infinities arise from integration at $s \sim 0$. Taylor expansion (26) and (31) are useful. For fictitious particles we obtain:

$$\ln\det\hat{F}^f = 4\int dx^4\sqrt{g}\sum_{n=0}^{\infty}\int_0^{\infty}\frac{ds}{s(4\pi s)^2}e^{\frac{i\sigma}{2s}}(is)^n a_n(x,x). \tag{34}$$

We put $\exp(i\sigma/2s)|_0 = 1$. Only coefficients at a_0, a_1 and a_2 are divergent at zero [quartic, square and logarithmic divergencies correspondingly]. So in (34) we shall consider only first three terms:

$$\ln\det\hat{F}^f = 4\int dx^4\sqrt{g}\int_0^{\infty}\frac{ds}{s(4\pi s)^2}e^{\frac{i\sigma}{2s}}(1 + isa_1 - s^2a_2)$$

$$= 4\int\frac{d^4x\sqrt{g}}{(4\pi)^2}\left(\frac{-4}{(\sigma+i0)^2} - \frac{2a_1}{\sigma+i0} + \left[\ln\frac{\sigma+i0}{2} - \int_0^{\infty}\frac{ds}{s}e^{i/s}\right]a_2\right). \tag{35}$$

Last integral is divergent at 0 and ∞. We can get similar expression for gravitons:

$$\ln \det \hat{F}^{\mathrm{Gr}} = \frac{1}{(4\pi)^2} \int d^4x \sqrt{g} \, \mathrm{tr}$$

$$\times \left(\frac{-4}{(\sigma + i0)^2} I - \frac{2\hat{a}_1}{\sigma + i0} + \left[\ln \frac{\sigma + i0}{2} - \int_0^\infty \frac{ds}{s} e^{i/s} \right] \hat{a}_2 \right) \Bigg|_{x=y}. \tag{36}$$

So we need coefficients a_1 and a_2 for both gravitons and fictitious particles. Then we can use (19) to calculate divergencies of the generating functional for scattering matrix. We can use (28) to derive:

$$\lim_{x \to y} a_n^f = \frac{1}{4n} \lim_{x \to y} \Delta^{-1/2} g^\alpha_{\beta'} (g^{\beta'}_\alpha \Delta^{1/2} a^f_{n-1})^\theta_{.\theta}. \tag{37}$$

Because the $\lim_{x \to y} \sigma^\mu = 0$. Equation (32) leads to:

$$\lim (a_n)^{\mu' \nu'}_{\alpha' \beta'} = \frac{1}{n} \lim_{x \to y} g^{\mu'}_\omega g^{\nu'}_\delta (\Delta^{1/2} g^\omega_{\lambda'} g^\delta_{\nu'} (a_{n-1})^{\lambda' \nu'}_{\alpha' \beta'})^\theta_{.\theta} - 2R^{\mu' \nu'}_{\lambda' \gamma'} (a_{n-1})^{\lambda' \gamma'}_{\alpha' \beta'}. \tag{38}$$

In order to calculate the right hand side in these equations we need to know expressions like $\sigma_{.\mu\nu\gamma\delta}$, which we can find recursively from equations: $\sigma^\mu \sigma_{.\mu} = 2\sigma$ also $\sigma^\mu g^\alpha_{\beta'.\mu} = 0$ and $4\Delta^{1/2} = 2\Delta^{1/2 \mu} \sigma_{.\mu} + \Delta^{1/2} \sigma^\mu_{.\mu}$. Last equation follows from (24). Now shall evaluate covariant derivatives and use commutation rule:

$$(\phi^\mu)_{.\nu\sigma} - (\phi^\mu)_{.\sigma\nu} = R_{\nu\sigma\tau}{}^\mu \phi^\tau. \tag{39}$$

Let us present a table of limits:

$$\lim \sigma = \lim \sigma^\mu_. = 0, \quad \lim \sigma_{.\mu\nu} = g_{\mu\nu}, \quad \lim \sigma_{.\alpha\beta\gamma} = 0,$$

$$\lim \sigma_{.\nu\sigma\tau\rho} = \frac{1}{3}(R_{\nu\tau\sigma\rho} + R_{\nu\rho\sigma\tau}) \tag{40}$$

$$\lim \sigma^{\mu\nu\sigma}_{.\mu\nu\sigma} = \frac{8}{5} R^\mu_{.\mu} + \frac{4}{15} R_{\mu\nu} R^{\mu\nu} - \frac{4}{15} R_{\alpha\beta\gamma\delta} R^{\alpha\beta\gamma\delta} \tag{41}$$

$$\lim \Delta^{1/2} = 1, \quad \lim (\Delta^{1/2})_{.\mu} = 0, \quad \lim \Delta^{1/2}_{.\mu\nu} = -\frac{1}{6} R_{\mu\nu},$$

$$\lim (\Delta^{1/2})^\nu_{.\mu\nu} = -\frac{1}{6} R_{.\mu} \tag{42}$$

$$\lim (\Delta^{1/2})^{\mu\nu}_{.\mu\nu} = -\frac{1}{5} R^\mu_{.\mu} + \frac{1}{36} R^2 - \frac{1}{30} R_{\mu\nu} R^{\mu\nu} + \frac{1}{30} R_{\alpha\beta\gamma\delta} R^{\alpha\beta\gamma\delta} \tag{43}$$

These limits were evaluated by De Witt in Ref. 6.

$$\lim g^\mu_{\nu'} = \delta^\mu_{\nu'}, \quad \lim g^\mu_{\nu'.\tau} = 0, \quad \lim g^\mu_{\nu'.\rho\lambda} = \frac{1}{2} R^\mu_{\nu' \rho\lambda} \tag{44}$$

$$\lim g^\mu_{\nu'.\rho\lambda\alpha} = \frac{1}{3} (R^\mu_{\rho\lambda\nu'.\alpha} + R^\mu_{\rho\lambda\nu'.\lambda}),$$

$$\lim g^{\mu\rho\alpha}_{\nu'.\rho\alpha} = \frac{1}{2} (R^{\mu\alpha\beta}_{\alpha\beta\nu'.} - R^\mu_{\alpha\beta\gamma} R^{\alpha\beta\gamma}_{\nu'}) \tag{45}$$

Using these tables and formula (37) we calculate the coefficients

$$a_0^f = 1, \quad \lim a_1^f = \frac{1}{4}\delta_\alpha^{\beta'}(\Delta^{1/2}g_{\beta'}^\alpha)_{.\mu}^\mu = -\frac{1}{6}R = 0 \tag{46}$$

$$\lim a_2^f = \frac{1}{32}\lim \delta_\alpha^{\beta'}[\Delta^{1/2}g_{\beta'}^\alpha g_\gamma^{\delta'}(g_{\delta'}^\gamma \Delta^{1/2})_{.\iota}^\iota]_{.\mu}^\mu$$

$$= -\frac{1}{10}R_{.\mu}^\mu + \frac{1}{72}R^2 - \frac{1}{60}R_{\mu\nu}R^{\mu\nu} - \frac{11}{240}R_{\alpha\beta\gamma\delta}R^{\alpha\beta\gamma\delta} \tag{47}$$

$$\lim a_2^f = -\frac{11}{240}R_{\alpha\beta\gamma\delta}R^{\alpha\beta\gamma\delta}. \tag{48}$$

Similar for gravitons:

$$\text{tr}\, a_0 = 9, \quad \text{tr}\, a_1 = 0, \quad \text{tr}\, a_2 = \frac{21}{40}R_{\alpha\beta\gamma\delta}R^{\alpha\beta\gamma\delta}. \tag{49}$$

Quadratic divergencies are absent both for gravitons and fictitious particles: $a_1^f = 0$ and $a_1^{\text{Gr}} = 0$ As for logarithmic divergencies, we should take into account the identity[6]:

$$\int d^4x\sqrt{g}(R^2 - 4R_{\mu\nu}R^{\mu\nu} + R_{\alpha\beta\gamma\delta}R^{\alpha\beta\gamma\delta}) = 0. \tag{50}$$

This means that:

$$\int d^4x\sqrt{g}(R_{\alpha\beta\gamma\delta}R^{\alpha\beta\gamma\delta}) = 0.$$

This means that logarithmic divergencies are also absent. So we proved that there is no ultra-violet divergencies for generating functional of scattering matrix on mass shell. This result was obtained simultaneously with Ref. 3.

4. Finite Part of the Generating Functional of the Scattering Matrix

We can rewrite Eq. (26) for Fourier transform:

$$f^f(x, x|s) = \int_{-\infty}^{\infty} d\omega e^{is\omega} f^f(x, x|\omega)$$

$$\frac{\partial(\omega f^f)}{\partial\omega} = \sigma^\mu f_{.\mu}^f + \frac{1}{4}g_{\beta'}^\alpha \Delta^{-1/2}\left(\nabla^\mu\nabla_\mu\Delta^{1/2}g_\alpha^{\beta'}\frac{\partial f^f}{\partial\omega}\right). \tag{51}$$

Similarly for gravitons we can define Fourier transform of f^{Gr}:

$$f_{\alpha'\beta'}^{\mu'\nu'}(x, x|s) = \int_{-\infty}^{\infty} d\omega e^{is\omega} f_{\alpha'\beta'}^{\mu'\nu'}(x, x|\omega).$$

We can start from Eq. (30) and obtain:

$$\frac{\partial(\omega f^{\mu'\nu'}_{\alpha'\beta'})}{\partial\omega} = \sigma^{\lambda}_{\cdot}(f_{\cdot\lambda})^{\mu'\nu'}_{\alpha'\beta'} + g^{\mu'}_{\omega}g^{\nu'}_{\delta}\Delta^{-1/2}\left(\Delta^{1/2}g^{\omega}_{\lambda'}g^{\delta}_{\nu'}\frac{\partial f^{\lambda'\nu'}_{\alpha'\beta'}}{\partial\omega}\right)^{\theta}_{.\theta}$$

$$-2g^{\mu'}_{\theta}g^{\nu'}_{\varsigma}R^{\theta\varsigma}_{\omega\delta}g^{\omega}_{\lambda'}g^{\delta}_{\gamma'}\frac{\partial f^{\lambda'\gamma'}_{\alpha'\beta'}}{\partial\omega}. \tag{52}$$

These function are used in diploma[2] to represent the finite part of the generating functional of the scattering matrix

$$iW = \frac{1}{2(4\pi)^2}\int d^4x\sqrt{g}\int_{-\infty}^{\infty}d\omega\left\{4f^f(x,x|\omega) - \frac{1}{2}\mathrm{tr}\,f^{\mathrm{Gr}}(x,x|\omega)\right\}\omega^2\ln\frac{\omega+i0}{m^2}. \tag{53}$$

Here m^2 is an arbitrary positive constant. The equations for f should be solved by perturbations starting from the flat metric.

$$iW = \frac{i\pi}{2(4\pi)^2}\int d^4x\sqrt{g}\int_{o}^{\infty}d\omega\left\{4f^f(x,x|\omega) - \frac{1}{2}\mathrm{tr}f^{\mathrm{Gr}}(x,x|\omega)\right\}\omega^2 \tag{54}$$

5. One Loop Diagram with Two Vertices

The section of diploma consider insertions of one loop diagram with one and two vertices in a tree diagram. The calculations in the diploma proves that the insertion of the diagram with one and two loops in any tree diagram vanish. The full text of diploma [in Russian] can be found on authors web-page.[2]

References

1. R. P. Feynman, R. B. Morinigo and W. G. Wagner, *Feynman Lectures on Gravitation*, ed. B. Hatfield (Addison-Wesley, 1995).
2. http://insti.physics.sunysb.edu/korepin
3. G. 't Hooft and M. Veltman, *Ann. Inst. Henri Poincare* **20**, 69 (1974).
4. L. D. Faddeev, Lectures in Leningrad State University.
5. N. P. Konopleva and V. N. Popov, *Gauge Fields* (Atomizdat, 1972).
6. B. S. De Witt, Dynamical theory of groups and fields, in *Relativity Groups and Topology* (Gordon and Breach, London, 1964).
7. P. A. M. Dirac, *Proc. Roy. Soc. A* **246**, 333 (1958).
8. R. Arnowitt, S. Deser and C. M. Misner, *Phys. Rev.* **117**, 1595 (1960).
9. J. Schwinger, *Phys. Rev.* **130**, 1253 (1963).
10. J. Schwinger, *Phys. Rev.* **132**, 1317 (1963).
11. P. Bergman, *Rev. Mod. Phys.* **33**, 510 (1961).
12. J. L. Anderson, *Rev. Mod. Phys.* **36**, 929 (1964)
13. L. D. Faddeev, Hamiltonian formulation of gravity, in *Proc Int. Conf. Gravity and Relativity* (Tbilisi, 1966).
14. S. N. Gupta, *Proc. Roy. Soc. A* **65**, 161 (1952).
15. S. N. Gupta, *Proc. Roy. Soc. A* **65**, 608 (1952).
16. R. P. Feynman, *Acta Phys. Polon.* **246**, (1963).

17. B. S. De-Witt, *Phys. Rev.* **160**, 1113 (1967).
18. B. S. De-Witt, *Phys. Rev.* **162**, 1195 (1967).
19. B. S. De-Witt, *Phys. Rev.* **162**, 1239 (1967).
20. V. N. Popov and L. D. Faddeev, *Phys. Lett.* B **25**, 29 (1967).

Chapter 2

QUANTUM DISCORD IN A SPIN SYSTEM WITH SYMMETRY BREAKING

BRUNO TOMASELLO[*,†,‡], DAVIDE ROSSINI[§],

ALIOSCIA HAMMA[¶] and LUIGI AMICO[*]

MATIS-INFM-CNR & Dipartimento di Fisica e Astronomia,
95123 Catania, Italy
†*SEPnet and Hubbard Theory Consortium, University of Kent,*
Canterbury CT2 7NH, UK
‡*ISIS facility, STFC Rutherford Appleton Laboratory,*
Harwell Oxford Campus, Didcot OX11 0QX, UK
§*NEST, Scuola Normale Superiore & Istituto Nanoscienze-CNR,*
Piazza dei Cavalieri 7, I-56126 Pisa, Italy
¶*Perimeter Institute for Theoretical Physics, 31 Caroline St. N,*
Waterloo ON, N2L 2Y5, Canada
† *bruno.tomasello@stfc.ac.uk*

We analyze the *quantum discord* Q throughout the low temperature phase diagram of the quantum XY model in transverse field. We first focus on the $T = 0$ order–disorder quantum phase transition QPT both in the symmetric ground state and in the symmetry broken one. Beside it, we highlight how Q displays clear anomalies also at a noncritical value of the control parameter inside the ordered phase, where the ground state is completely factorized. We evidence how the phenomenon is in fact of collective nature and displays universal features. We also study Q at finite temperature. We show that, close to the QPT, Q exhibits quantum-classical crossover of the system with universal scaling behavior. We evidence a nontrivial pattern of thermal correlations resulting from the factorization phenomenon.

Keywords: Quantum phase transitions; quantum information; quantum correlations.

1. Introduction

Correlations provide a characterization of many-body systems.[1] In the quantum realm, beside classical correlations, nonlocal quantum correlations (like *entanglement*) play a pivotal role. Although entanglement completely describes quantum correlation for *pure* states, it is in general more subtle to characterize the pattern of correlations for *mixed* states. Indeed the quantitative interplay between classical and quantum correlations has been formulated only recently with the introduction of the *quantum discord*, operatively defining pure quantum correlations in composite systems.[2,3,5–7]

The phase diagram of spin systems displays nontrivial pattern of correlations dictated by two main features: the quantum critical point (QCP) and the factorizing point. In fact at zero temperature the system can undergo an order–disorder Quantum Phase Transition (QPT), as long as the control parameter h is tuned across a critical value h_c.[8] It is worth noting that the quantum order arises because superselection rules lead to a symmetry breaking.[4] Besides QPT, spin systems may display a further remarkable phenomenon occurring at $h = h_f$ located within the ordered symmetry broken phase, where the ground state is exactly factorized,[9–13] and therefore correlations are exclusively classical. Such factorization consists in a transition for the two-spin-entanglement[15] and is rigorously not accompanied by any change of symmetry.

In this article we analyze the quantum discord arising of the quantum XY spin system both at zero and finite temperature. In particular we consider the ground state with broken symmetry. We show that, besides the usual critical behavior at the QPT, the quantum discord displays dramatic changes also at the factorizing point, within the ordered phase (i.e., with nonvanishing spontaneous magnetization). We complete our study by detecting how the quantum critical and the factorization point affect the quantum discord at low-temperature, thus opening the way towards actual observations. The structure of this paper is as follows. In the first section an overview is given about the current notions of quantum and classical correlation in a general quantum system. In the second one we introduce a many body system suitable for our type of analysis; the Hamiltonian of the model is introduced together with few fundamental features related to its physics. In the third section we show our analysis and results at zero temperature and then once the temperature is switched on.

2. Quantum, Classical, and Total Correlations

In a bipartite system AB the total amount of correlations between A and B is given by the mutual information

$$I(A : B) \equiv S(\hat{\rho}^A) + S(\hat{\rho}^B) - S(\hat{\rho}^{AB}), \tag{1}$$

where $S(\hat{\rho}) = -\text{Tr}[\hat{\rho} \ln \hat{\rho}]$ is the von Neumann entropy. In the classical information, using the *Bayes rule*, an equivalent formulation of mutual information is:

$$J(A : B) \equiv S(A) - S(A|B), \tag{2}$$

where the conditional entropy $S(A|B) = S(AB) - S(B)$ quantifies the ignorance on part A once a measurement on B is performed. But in the quantum realm a measurement in general perturbs the system and part of the information itself is lost. So when we consider a quantum composite system the Eq. (2) differs from Eq. (1). This *difference* allow us to estimate the relative role of quantum and classical correlations in quantum composite systems.[2,3]

Indeed if we describe a measurement on part B by a set of projectors $\{\hat{B}_k\}$, then

$$\hat{\rho}^{AB}_{(k)} = \frac{1}{p_k}(\hat{\mathbb{I}}_A \otimes \hat{B}_k)\hat{\rho}^{AB}(\hat{\mathbb{I}}_A \otimes \hat{B}_k) \tag{3}$$

is the composite state conditioned to the kth outcome with probability $p_k = \text{Tr}[(\hat{\mathbb{I}}_A \otimes \hat{B}_k)\hat{\rho}^{AB}(\hat{\mathbb{I}}_A \otimes \hat{B}_k)]$. This conditioned state is the key ingredient which distinguishes between classical and quantum correlations: in fact it differs in general from the pre-measurement state $\hat{\rho}^{AB}$ as well as mutual information differs from Classical correlations. Then the amount of classical correlations C is obtained by finding the set of measurements on $\{\hat{B}_k\}$ that disturbs the least the part A, i.e., by maximizing $C = \max_{\{\hat{B}_k\}}[S(\hat{\rho}^A) - S(\hat{\rho}^{AB}|\{\hat{B}_k\})]$.[2,3,5–7] Then the difference between mutual information and Classical correlations defines the so-called *Quantum Discord*

$$Q(\hat{\rho}^{AB}) \equiv I(A:B) - C(\hat{\rho}^{AB}). \tag{4}$$

In the estimate of quantum correlations between subsystem of a bipartite system the Entanglement has been playing a leading role, in particular about the relevance of correlations in many body systems. However Quantum Discord differs in general from Entanglement: for example, even if they are the same for pure states, they can display a very different behavior in mixed states.

3. The XY Model in Transverse Field

We will consider here an interacting pair of spin-1/2 in the anti-ferromagnetic XY chain with transverse field h. The Hamiltonian of the model

$$\hat{\mathcal{H}} = -\sum_j \left(\frac{1+\gamma}{2}\hat{\sigma}^x_j\hat{\sigma}^x_{j+1} + \frac{1-\gamma}{2}\hat{\sigma}^y_j\hat{\sigma}^y_{j+1} + h\hat{\sigma}^z_j\right), \tag{5}$$

describes the competition between two parts: the anisotropy on the xy plane (tuned by varying $\gamma \in (0,1)$ and the coupling with external magnetic field h along the z direction. Using a set of successive transformations (Jordan-Wigner, Bogoliubov, Fourier[20]), the Pauli matrices operators $\hat{\sigma}^\alpha_j$ ($\alpha = x, y, x$) on sites j can be expressed in terms of operators such that the Hamiltonian takes the diagonal form $\hat{\mathcal{H}} = -\sum_k \Lambda_k \eta^\dagger_k \eta_k + \text{const}$. Here the system is described as a gas of noninteracting fermions, where $\eta^\dagger_k$ (η_k) is the creation (annihilation) operator of a fermion with momentum k. Furthermore the Jordan–Wiegner transformations allows an *analytic* expression for the correlation functions $g_{\alpha\alpha}(r)$ of any two spins in the chain far r sites with each other (because of translational invariance the distance between them is all that matters).[18,19] In fact the exact solution of XY model has encouraged extensively studying on the critical phenomena it displays.[17–19] In particular, during the last decade through the analysis of quantum correlations (i.e., *Entanglement*) new insights has been made in the description of the physics of the system.[21]

3.1. *The Phase Diagram*

The phase diagram of the XY model is characterized by *two* values of the applied field h.[8–10,21] It is well-known indeed that for $\gamma \in (0,1]$ the system displays a continuum QPT for $T = 0$, $h_c = 1$, of the Ising universality class with critical indices $\nu = z = 1$, $\beta = 1/8$.[8] In fact, for strong enough external fields ($h \gg h_c$) all spins tends to be aligned along the z direction, while the opposite limit ($h \ll h_c$) give rise to a spontaneous magnetization (Z_2-symmetry broken) along a direction on the xy plane, γ dependent. Then at zero temperature on the left side $h < h_c$ of the phase diagram the system is an ordered ferromagnet and the Z_2-symmetry is broken, while on the right side $h > h_c$ the quantum fluctuations leads to the disordered phase and the system is a quantum paramagnet.

At finite temperature the physics of the whole system is affected from the QCP $h = h_c$ at zero temperature. A V-shaped diagram in the $h - T$ plane emerges, characterized by the straight lines $T = |h - h_c|$ that mark the crossover region between the so-called *Quantum Critical Region* ($T > |h - h_c|$) and the *Quasi Classical* regions surrounding it.[8]

Besides the QCP h_c there's another value of the transverse field that characterize the phase diagram of the XY model. In fact at zero temperature, given a certain anisotropy γ, there is one specific value $h_f = \sqrt{1 - \gamma^2}$ where the ground state is exactly factorized[9,10]

$$|\Psi^\gamma_{GS}\rangle = \prod_j |\psi^\gamma_j\rangle . \tag{6}$$

At this particular value of field it seems that, even though the system is in a phase with very strong quantum correlations, there is a "critical" set of values $h_f(\gamma)$ where the state is completely classical. This strange occurrence, regarded as a paradox in the first place,[9,10] seems to be strongly connected with the reshuffling of correlations among the system. In fact, a deep analysis on the behavior of Entanglement has remarkably shed new light on the relevant physics involved on h_f.[11–13] In particular it has been shown that tuning the external field from $h < h_f$ to $h > h_f$ the entanglement pattern swaps from parallel to anti-parallel.[15] Furthermore, it has been observed that at zero temperature the bipartite entanglement has a logarithmically divergent range at h_f, together with the fact that at finite temperature there is a whole region fanning out from h_f where no pairwise entanglement survives.[16]

Then there is strong evidence, that along these critical values of field and Temperature, the behavior of Entanglement, and correlations in general, play a pivotal role in the physics involved and hence in our understanding of it. In particular it seems that the interplay of Correlations when the field is tuned across h_f is the only accessible way, so far, to tackle the puzzling physics that leads to the factorized state (6). In fact we found here that the Quantum Discord allows a fine structure of the phase diagram around h_c, and most important displays a nontrivial scaling law at the factorization field h_f.

4. Classical and Quantum Correlations in the Model

In order to compute Q_r between any two spins A and B at distance r along the chain is the key ingredients are density matrices. In fact the Eq. (4) depends both on the single site density matrices $\hat{\rho}^A$, $\hat{\rho}^B$ and on the two sites density matrix of the composite subsystem $\hat{\rho}^{AB}$. Due to the translational invariance along the chain, the single site density matrix is the same for any spin (dependent only on $m_z = \langle \sigma_z \rangle / 2$), in particular

$$
\hat{\rho}^A = \hat{\rho}^B = \begin{pmatrix} \dfrac{1}{2} + m_z & 0 \\ 0 & \dfrac{1}{2} - m_z \end{pmatrix}. \tag{7}
$$

Then the single site von Neumann entropy for both spin A and spin B is:

$$
S(\hat{\rho}^A) = S(\hat{\rho}^B) = S_{\text{bin}} \left(\frac{1}{2} + m_z \right) \tag{8}
$$

where $S_{\text{bin}}(p) = -p \log p - (1-p) \log(1-p)$ is the binary Shannon entropy.

On the other hand the expression of $\hat{\rho}^{AB}$ may be cumbersome. In fact, the general 2 sites reduced density matrix for an Hamiltonian model with global phase flip symmetry has the following form[27–29]:

$$
\hat{\rho}_r = \begin{pmatrix} A & a & a & F \\ a & B & C & b \\ a & C & B & b \\ F & b & b & D \end{pmatrix} \tag{9}
$$

in the basis $\{|00\rangle, |01\rangle, |10\rangle, |11\rangle\}$, where $|0\rangle$ and $|1\rangle$ are eigenstates of σ^z, and we'll shortly see explicitly the matrix elements.

Because of translational invariance, this density matrix depends only on the distance r between the two spins, $\hat{\rho}^{AB} = \hat{\rho}_r$. In particular note that A and B in the matrix (9) do not denote the two spins considered, but some of the following quantities related to the correlators $g_{\alpha\beta}(r) = \langle \hat{\sigma}_j^\alpha \hat{\sigma}_{j+r}^\beta \rangle$ and $g_\alpha = \langle \sigma_\alpha \rangle = 2m_\alpha$:

$$
A = \frac{1}{4}(1 + g_z + g_{zz}),
$$

$$
D = \frac{1}{4}(1 - g_z + g_{zz}),
$$

$$
B = \frac{1}{4}(1 - g_{zz}), \tag{10}
$$

$$
C = \frac{1}{4}(g_{xx} + g_{yy}),
$$

$$
F = \frac{1}{4}(g_{xx} - g_{yy}).
$$

are the parity coefficients, while

$$a = \frac{1}{4}(g_x + g_{xz}),$$

$$b = \frac{1}{4}(g_x - g_{xz}),$$

(11)

explicit the contribution from the symmetry breaking.

As long as the system is in the Z_2-symmetric phase the matrix element in "low case" are null ($a = b = 0$). The symmetry breaking manifest itself in a, $b \neq 0$.[27-29] In the former case the remaining nonvanishing e entries in Eq. (9) can be evaluated analytically,[18,19] and we use a fully analytical approach to compute the Quantum Discord in the so-called *thermal ground state*.[33] In this state the system approach the ground state by lowering the temperature towards the limit of $T = 0$, but for this reason the symmetry is conserved and the state is not in the "true" degenerate *ground state*. In the latter one the Z_2-symmetry is lost and beside the spontaneous magnetization m_x, also the nontrivial $g_{xz}(r)$ appears.[30] In this case we analyze the real ground state by means of *numerical* methods, i.e., Density Matrix Renormalization Group (DMRG) for finite systems with open boundaries.[31] Once we have access to the density matrices through the correlation function, we need to compute the explicit form of the mutual information and the classical correlation in order to "distill" the amount of pure quantum correlations [i.e., Quantum Discord, Eq. (4)]. Here we follow the notation used in Ref. 33. Since the reduced density matrix of the single spin is the same for any site, we have already shown in Eq. (8) that $S(\hat{\rho}^A) = S(\hat{\rho}^B)$. Hence the mutual information is:

$$I(\hat{\rho}_r^{AB}) = 2S(\hat{\rho}^A) - \sum_{\nu=0}^{3} \lambda_\nu \log \lambda_\nu$$

(12)

where $\lambda_\nu(r)$ are eigenvalues of $\hat{\rho}_r^{AB}$, that in terms of correlation functions $g_{\alpha\alpha}(r) = \langle \hat{\sigma}_j^\alpha \hat{\sigma}^\alpha j + r \rangle$ and $g_z = 2m_z$ are[33]:

$$\lambda_0 = \frac{1}{4}\left(1 + g_{zz} + \sqrt{g_z^2 + (g_{xx} - g_{yy})^2}\right)$$

$$\lambda_1 = \frac{1}{4}\left(1 + g_{zz} - \sqrt{g_z^2 + (g_{xx} - g_{yy})^2}\right)$$

(13)

$$\lambda_2 = \frac{1}{4}\left(1 - g_{zz} + (g_{xx} + g_{yy})\right)$$

$$\lambda_2 = \frac{1}{4}\left(1 - g_{zz} - (g_{xx} + g_{yy})\right).$$

Once the mutual information is known in terms of the correlation functions [Eq. (12)] we need to find the explicit form for the classical correlations in the *XY* model, in order to compute pure quantum correlations, as stated in Eq. (4). Following a procedure similar to Refs. 32–34, we use a set of projectors $\{\hat{B}_k\}$ as local measurements on the spin B. In particular, working on the computational basis $\{|0\rangle, |1\rangle\}$ in the Hilbert space $\mathcal{H}_B^2$ associated to the spin B, our general set of projectors is:

$$\{\hat{B}_k = V\hat{\Pi}_k V^\dagger\}, \quad k = 0, 1 \tag{14}$$

where $\hat{\Pi}_k = |k\rangle\langle k|$ related to the basis vectors and $V \in U(2)$ gives the generalization to any type of projector on B. As suggested in Ref. 33, it is useful to parametrize V as follows:

$$V = \begin{pmatrix} \cos\dfrac{\theta}{2} & \sin\dfrac{\theta}{2}e^{-i\phi} \\ \sin\dfrac{\theta}{2}e^{i\phi} & -\cos\dfrac{\theta}{2} \end{pmatrix} \tag{15}$$

where $\theta \in [0, \pi]$ and $\phi \in [0, 2\pi]$ are respectively the azimuthal and polar axes of a qubit over the *Bloch sphere* in $\mathcal{H}_B^2$. After a measurement has been performed on B the reduced density matrix $\hat{\rho}_{\{\hat{B}_k\}}^{AB}$ will be in one of the following states:

$$\hat{\rho}_0^{AB} = \frac{1}{2}\left(\hat{\mathbb{I}}_A + \sum_{\alpha=1}^{3} q_{0\alpha}\hat{\sigma}_A^\alpha\right) \otimes (V\hat{\Pi}_0 V^\dagger)_B,$$

$$\hat{\rho}_1^{AB} = \frac{1}{2}\left(\hat{\mathbb{I}}_A + \sum_{\alpha=1}^{3} q_{1\alpha}\hat{\sigma}_A^\alpha\right) \otimes (V\hat{\Pi}_1 V^\dagger)_B. \tag{16}$$

This expression for the reduced density matrices gives the explicit dependence of the system A respect to the projective measurement performed on the spin B. Indeed the coefficient $q_{k\alpha} = q_{k\alpha}(\theta, \phi)$ in the expansion depend on the projectors used to perform the measure on B (see Ref. 33 for the explicit form).

We remind here the explicit form for the classical correlations[2,3,5–7]

$$C = \max_{\{\hat{B}_k\}}[S(\hat{\rho}^A) - S(\hat{\rho}^{AB}|\{\hat{B}_k\})]. \tag{17}$$

Maximizing over all possible $\hat{B}_k$ is equivalent to find for those values (θ, ϕ) that *disturb* the least the spin A when we make a measure on B. We found $(\theta = \pi/2, \phi = 0)$, in agree with Refs. 33 and 35. And following their same method to evaluate $S(\hat{\rho}^{AB}|\{\hat{B}_k\})$ we found a simple expression for classical correlations

$$C_r = H_{\mathrm{bin}}(p_1) - H_{\mathrm{bin}}(p_2), \tag{18}$$

where $H_{\mathrm{bin}}(p)$ is the Shannon entropy and

$$p_1 = \frac{1}{2} + m_z$$

$$p_2 = \frac{1}{2} + \sqrt{g_{xx}^2/4 + m_z^2}. \tag{19}$$

So by the difference of the mutual information (12) and the latter expression for the classical correlations (18) we get the simplified expression for the quantum discord between any two spins in the XY chain in transverse field

$$Q(\hat{\rho}_r) = H_{\text{bin}}(p_1) + H_{\text{bin}}(p_2) - \sum_{\nu=0}^{4} \lambda_\nu \log \lambda_\nu . \qquad (20)$$

5. Analysis of Quantum Discord at $T = 0$

In this section, we show our analysis of quantum correlations both in the thermal ground state and in the symmetry broken one. In particular we remark differences and similarities between them, and highlight the interesting features occurring at the QCP $h_c = 1$ and at the factorizing field h_f (if not specified $\gamma = 0.7$ in every picture).

We start by showing the behavior of quantum discord at zero temperature, over a wide range of external field h centered around the critical value h_c where the QPT occurs. In Fig. 1 we compare, for both the XY model ($\gamma = 0.7$ for example) and the Ising one (inset where $\gamma = 1$), the numerical results we got from DMRG

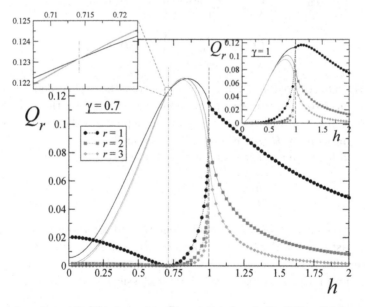

Fig. 1.   Quantum discord $Q_r(h)$ between two spins at distance r in the XY model at $\gamma = 0.7$ (main plot and left inset) and $\gamma = 1$ (right inset), as a function of the field h. Continuous lines are for the thermal ground state, while symbols denote the symmetry-broken state obtained by adding a small symmetry-breaking longitudinal field $h_x = 10^{-6}$ and it was computed with DMRG in a chain of $L = 400$ spins; simulations were performed by keeping $m = 500$ states and evaluating correlators at the center of the open-bounded chain. For $\gamma = 0.7$ and at $h_f \simeq 0.714$, in the symmetric state all the curves for different values of r intersect, while after breaking the symmetry Q_r is rigorously zero. At the critical point Q_r is nonanalytic, thus signaling the QPT. In the paramagnetic phase, there is no symmetry breaking to affect Q_r.

computation of Q_r on the true ground state respect to the analytical values of the thermal ground state. There is strong evidence that in the disordered phase, $h > h_c$, no difference occurs between the two state, while in the opposite regime $h < h_c$ two different pattern comes out. In fact in the latter case, the order in the system is really achieved only in the case where the symmetry of the system is lost. We achieve this condition using a staggered field in the DMRG computation that leads to the symmetry breaking and gives out the true ground state where quantum correlations are very small as long as $h < h_c$. In particular it is remarkable that they start to increase once the field is tuned immediately upper the factorizing field, where all quantum correlations must vanish, to reach a cuspid-like maximum at the QCP. On the other hand, the quantum discord on the thermal ground state (solid lines in Fig. 1) is a smooth function respect to the field. In general it depends on the distance r between the two spins, but at the factorizing point it gets the same value for any length scale.[36]

To go deeper in the analysis let us focus on $h_c = 1$ in the first place. The QPT is in general marked by a divergent derivative of the quantum discord, with respect to the field. In particular such divergence is present for any γ in the symmetry broken state, while on the thermal ground state it holds as long as $\gamma < 1$ (see Fig. 2); for $\gamma = 1$, $\partial_h Q_r$ is finite at h_c although the $\partial_h^2 Q_r$ diverges.[33] This divergence suggests that a scaling analysis at the QCP is feasible. In particular in Fig. 3 we show the finite size scaling $\partial_h Q_{r=1}$ for the symmetry-broken ground state in proximity of h_c. We found that $z = \nu = 1$, thus meaning that the transition is in the Ising universality class.

Turning now into the factorizing field we underline that, for the thermal ground state, it is the only value where the curves with different r, intersect with each other

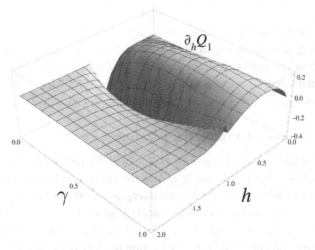

Fig. 2. Behavior of $\partial_h Q_r$ in the thermal ground state, with respect to the field for any type of anisotropy γ. Focusing at the QCP ($h = 1$), note that for $\gamma < 1$ it is divergent, while it remains finite for $\gamma = 1$.

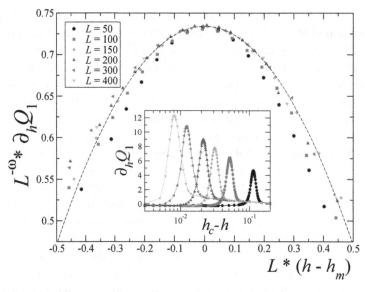

Fig. 3.   Finite-size scaling of $\partial_h Q_1$ for the symmetry-broken state in proximity of the critical point h_c. Displayed data are for $\gamma = 0.7$. The first derivative of the quantum discord is a function of $L^{-\nu}(h - h_m)$ only, and satisfies the scaling ansatz $\partial_h Q_1 \sim L^\omega \times F[L^{-\nu}(h - h_m)]$, where h_m is the renormalized critical point at finite size L and $\omega = 0.472$. We found a universal behavior $h_c - h_m \sim L^{-1.28\pm0.03}$ with respect to γ. Inset: raw data of $\partial_h Q_1$ as a function of the transverse field.

(see up-left inset in Fig. 1).[36] Beside this, in the broken symmetry state, not only we found that all curves vanish in h_f, but we numerically estimated the following particular dependence of Q_r close to it:

$$Q_r \sim (h - h_f)^2 \times \left(\frac{1-\gamma}{1+\gamma}\right)^r . \qquad (21)$$

Such behavior is consistent with the expression of correlation functions close to the factorizing line obtained in Ref. 14, and here appears to incorporate the effect arising from the nonvanishing spontaneous magnetization. Most remarkably, we found a rather peculiar dependence of Q_r on the system size, converging to the asymptotic value $Q_r^{(L\to\infty)}$ with an exponential scaling behavior (see Fig. 4).

6. Quantum Discord at Finite Temperature

Even if both the QCP h_c and the factorizing field h_f are defined at $T = 0$, they influence the whole physics of the system once the temperature is switched on. Close to h_c, the physics is dictated by the interplay between thermal and quantum fluctuations of the order parameter. As we stated before the cross-over temperature $T_{\text{cross}} = |h - h_c|^z$ fixes the energy scale.[8] For $T \ll T_{\text{cross}}$ the system is described by a quasi-classical theory, while inside the "quantum critical region" ($T \gg T_{\text{cross}}$), it's impossible to distinguish between quantum and thermal effects. Here the critical property arising from the QCP at $T = 0$ are highly dominating the dynamics

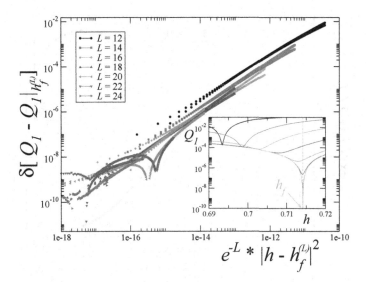

Fig. 4. Scaling of Q_1 close to the factorizing field, for $\gamma = 0.7$: we found an exponential convergence to the thermodynamic limit, with a universal behavior according to $e^{-\alpha L}(h - h_f^{(L)})$, $\alpha \approx 1$ [$h_f^{(L)}$ denotes the effective factorizing field at size L, while $\delta(Q_1) \equiv Q_1^{(L)} - Q_1^{(L \to \infty)}$]. Due to the extremely fast convergence to the asymptotic value, already at $L \sim 20$ differences with the thermodynamic limit are comparable with DMRG accuracy. Inset: raw data of Q_1 as a function of h. The cyan line is for $L = 30$ so that, up to numerical precision, the system behaves at the thermodynamic limit.

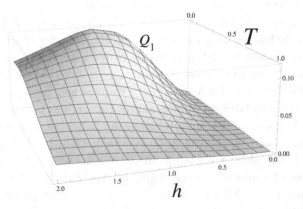

Fig. 5. Quantum discord in the thermal state of the Ising model with $\gamma = 1$, as a smooth function of temperature T and of the external field h.

of the system, and we would aspect that quantum correlations show some particular pattern as well as they do at h_c. In fact close to h_f and at small T, the bipartite entanglement remains vanishing in a finite nonlinear cone in the $h - T$ plane.[16,21] Thermal states, though, are not separable, and entanglement is present in a multipartite form.[26] In this regime the bipartite entanglement results to be nonmonotonous, and a reentrant swap between parallel and antiparallel entangle-

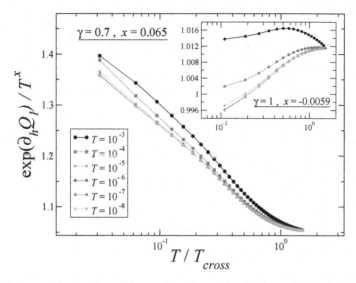

Fig. 6.   Finite-temperature scaling of the quantum discord for the thermal state close to the critical point. The logarithmic scaling is verified: along the critical line we found $\partial_h Q_1|_{h_c} \sim x \ln(T) + k$, with $x = 0.065$ for $\gamma = 0.7$. The scaling function F shows a data collapse close to the critical point. Inset: same analysis for the Ising case ($\gamma = 1$); we found an analogous scaling behavior with $x = -0.0059$.

ment is observed.[16] At finite temperature, the Z_2-Symmetry is preserved all over the values of h (there is no longer a symmetry broken phase). This means that if the system lies in the ground state at $T = 0$ (symbols lines in Fig. 1), once the temperature is switched on we get a jump of Q_r all along the phase $h < h_c$. After that it behaves as a smooth function decaying with temperature (Fig. 5). Such discontinuity is also observed in the entanglement, even if in that case it is much less pronounced and occurs only for $h < h_f$.[27–29] We now analyze how criticality and factorization modify the fabric of pure quantum correlations in the $h - T$ plane.

We start by focusing on the finite-temperature scaling of the quantum discord close to the critical point h_c. In the first place we verified the logarithmic scaling $\partial_h Q_r|_{h_c} \sim x \ln(T) + k$ along the critical line, $h = 1$ in the $h - T$ plane (see Fig. 9), where the value of x depends on the degree of anisotropy γ. Once x is given (for example we found $x = 0.065$ for $r = 1$, $\gamma = 0.7$, Fig. 6), by properly tuning the ratio T/T_{cross}, where $T_{\text{cross}} \equiv |h - h_c|$, we verified the scaling ansatz

$$\partial_h Q_r = T^x F\left(\frac{T}{T_{\text{cross}}}\right). \tag{22}$$

In particular in Fig. 6 we show how different curves, related to different values of T/T_{cross}, collapse when approaching the critical point. Remarkably in the Ising case (inset) the scaling is verified as well, even if the derivative $\partial_h Q_1$ is finite at h_c. To explore the behavior of correlations along the $h - T$ plane we studied how the quantum discord varies on the phase diagram just above the QCP. In the first place

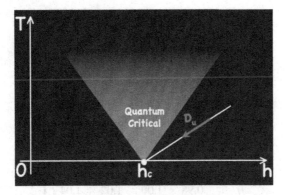

Fig. 7. Schematic representation for the directional derivative on the phase diagram. It allows to study how quantities vary along straight lines coming out from the critical point with slope $\mathbf{u} \equiv (\cos \alpha, \sin \alpha)$.

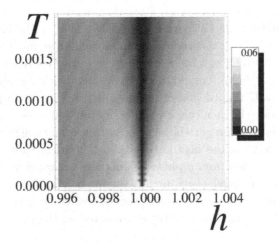

Fig. 8. Density plot in the $h - T$ plane of $D_{\mathbf{u}}Q_1$ close to h_c; The vertical line starting from the QCP shows that Q_1 tends to be constant inside the quantum critical region.

we analyze how the derivative respect to field behaves along the directions fanning out from h_c. In Fig. 7 we sketch a cartoon to describe the directional derivative $D_{\mathbf{u}}Q = |\partial_T Q \sin \alpha + \partial_a Q \cos \alpha|$ we used to describe how Q_1 varies close to the QCP. From the pattern of $D_{\mathbf{u}}Q_1$ at low temperature (Fig. 8) we see how the presence of the QPT charaterizes the whole phase diagram. The black vertical line starting from the QCP highlights the fact that the quantum discord remains constant along the critical line $h = 1$: in a sense, close to such region $h \approx 1$, quantum correlations are particularly "rigid". This explains their robustness up to finite temperatures, particularly along slopes within the quantum critical region. On the other hand, out of the quantum critical region, the variation of Q_1 is drastically increased. We also point out the peculiar asymmetric behavior between the two semiclassical regions (in the ordered phases $D_{\mathbf{u}}Q_1$ is generally higher than in the paramagnetic phase).

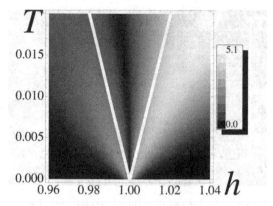

Fig. 9.   Density plot in the $h - T$ plane of $\partial_T[Q_1/C_1]$ close to h_c; along the critical line the ratio Q_1/C_1 is constant with respect to the temperature. The solid straight line $(T = |h - h_c|)$ marks the boundary of the quantum critical region.

Furthermore, to make a more accurate analysis we look at the interplay between quantum discord and classical correlations. In particular we analyze how this ratio Q_1/C_1 varies with the temperature, exploiting the respective sensitivity to thermal fluctuations arising at finite temperature. Accordingly with the phase diagram related to the QPT, a V shaped pattern comes out (see Fig. 9). In particular along the critical line, inside the quantum critical region, we found $\partial_T[Q_1/C_1] = 0$. Then apparently the ratio between correlations is constant even though the temperature is switched on, as long as the field is tuned at the critical value h_c. Beside this, the whole crossover region from a phase to another is marked as the highest variation in the nature of correlations in the system. In conclusion we analyze how the factorizing field affects the physics of the system at nonzero temperature. As we emphasized before, the Z_2-symmetry is preserved on the thermal state. In par-

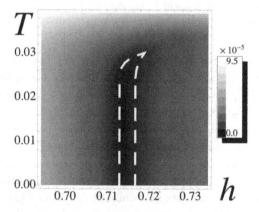

Fig. 10.   Average quantum discord displacement: $\overline{\Delta Q_r} = 2\sum_{i,j=1}^{m} |Q_{r_i} - Q_{r_j}|/m(m-1)$ for $m = 5$ fanning out from the factorizing point $h_f \sim 0.714$, where all correlations coincide at any length scale r, as evidenced in the left inset of Fig. 1. Here $\gamma = 0.7$.

ticular we highlighted also that for the thermal ground state, the factorizing field was the unique value where the quantum discord is the same at any length scale r. Here we found that indeed this feature is present even after the temperature is switched on. In fact in Fig. 10 we propose $\overline{\Delta Q_r} = 2 \sum_{i,j=1}^{m} |Q_{r_i} - Q_{r_j}|/m(m-1)$ as a measure of the robustness of this characteristic at nonzero temperature. We consider different distances between the couple of spin A and B, and we take the average of the difference between the respective quantum discord. There is strong evidence in Fig. 10 that for a finite range of temperature this difference is still zero (i.e., the quantum discord for different r is still the same only at h_f).

Here we emphasize that the robustness and sensitivity of the quantum discord to nonzero temperature, encourage the implementation of suitable experiments that could give good feedback of our analysis.

7. Discussions

We studied pure quantum correlations quantified by the quantum discord Q_r in the quantum phases involved in a symmetry-breaking QPT. In the ordered phase, although Q_r results relatively small in the symmetry broken state as compared to the thermal ground state, it underlies key features in driving both the order–disorder transition across the QPT at h_c, and the correlation transition across the factorizing field h_f. The critical point is characterized by a nonanalyticity of Q_r found in the Ising universality class. Close to h_f, Q_r displays uniquely nontrivial properties: in the thermal ground state quantum correlations are identical at all scales; for the symmetry broken state the factorization can be interpreted as a collective reshuffling of quantum correlations. We point out that h_f marks the transition between two "phases" characterized by a different pattern of entanglement.[15,16] Accordingly our data provide evidence that such a *correlation transition* phenomenon is of collective nature, governed by an exponential scaling law. We observe that the scaling close to h_f cannot be algebraic because the correlation functions decay exponentially in the gapped phases. We notice however that, due to the peculiar phenomenology of the factorizing phenomenon, it can be specific for scaling beyond the generic exponential behavior that is observed in gapped phases. For finite L different ground states for the two parity sectors intersect.[22,23] The ground state energy density is diverging *for all L* (such divergence, though, vanishes in the thermodynamic limit). Indeed we found that the factorization occurs without any violation of adiabatic continuity. Accordingly, the ground state fidelity $\mathcal{F}(h)$, which can detect both symmetry breaking and nonsymmetry breaking QPT, is a smooth function at h_f.[40] We remark that this can occur without closing a gap and changing the symmetry of the system, as a signature of the fact that quantum phases and entanglement are more subtle than what the symmetry-breaking paradigm says. Such a behavior is particularly relevant in the context of QPTs involving topologically ordered phases where a QPT consists in the change of the global pattern of entanglement, instead of symmetry.[37-39]

We analyze the phase diagram at low T. A discontinuity of Q_r with T is evidenced in the whole ordered phase $h < h_c$. We proved that Q_r displays universal features, identifying the quantum critical region, as that one where the quantum discord (relatively to classical correlation) is frozen out to the $T = 0$ value. In particular we notice that in each phase the ratio between correlations is stable respect to the temperature, while the highest variation are in particular in the crossover region on the right of h_c, where the system is running out of the critical region into the quasi-classical one just above the disordered phase of the paramagnet. This aspect shows that this type of quantum correlations allows a fine structure of the phase diagram according to the behavior of the gap $\Delta \lesssim 0$ in the low temperature limit $T \ll |\Delta|$, that takes into account that the mechanism leading to the two corresponding semiclassical regimes driven from quantum ($\Delta > 0$) or thermal ($\Delta < 0$) fluctuations.[8]

We have found that a nontrivial pattern of quantum correlations fans out from the factorization of the ground state as well.

Acknowledgments

We thank A. De Pasquale, R. Fazio, S. Montangero, D. Patané, M. Zannetti, J. Quintanilla for useful discussions. The DMRG code released within the PwP project (www.dmrg.it) has been used. Research at Perimeter Institute for Theoretical Physics is supported in part by the Government of Canada through NSERC and by the Province of Ontario through MRI. DR acknowledges support from EU through the project SOLID.

References

1. X.-G. Wen, *Quantum Field Theory of Many-Body Systems* (Oxford University Press, USA, 2004).
2. L. Henderson and V. Vedral, *J. Phys. A: Math. Gen.* **34**, 6899 (2001).
3. H. Ollivier and W. H. Zurek, *Phys. Rev. Lett.* **88**, 017901 (2002).
4. S. Coleman, Secret symmetry: An introduction to spontaneous symmetry breakdown and gauge fields, in *Laws of Hadronic Matter*, ed. A. Zichichi (Academic Press, New York, 1975).
5. V. Vedral, *Phys. Rev. Lett.* **90**, 050401 (2003).
6. D. Borivoce, V. Vedral and C. Bruckner, *Phys. Rev. Lett.* **105**, 190502 (2010).
7. A. Auyuanet and L. Davidovic, *Phys. Rev. A* **82**, 032112 (2010).
8. S. Sachdev, *Quantum Phase Transitions* (Cambridge University Press, Cambridge, 2001).
9. J. Kurmann, H. Thomas and G. Muller, *Physica A* **112**, 235 (1982).
10. T. Roscilde *et al.*, *Phys. Rev. Lett.* **94**, 147208 (2005).
11. S. M. Giampaolo *et al.*, *Phys. Rev. Lett.* **100**, 197201 (2008).
12. S. M. Giampaolo *et al.*, *Phys. Rev. B* **79**, 224434 (2009).
13. S. M. Giampaolo *et al.*, *Phys. Rev. Lett.* **104**, 207202 (2010).
14. F. Baroni *et al.*, *J. Phys. A: Math. Theor.* **40**, 9845 (2007).
15. A. Fubini *et al.*, *Eur. Phys. J. D* **38**, 563 (2006).
16. L. Amico *et al.*, *Phys. Rev. A* **74**, 022322 (2006).

17. P. Pfeuty, *Ann. Phys.* **57**, 79 (1970).
18. E. Barouch and B. M. McCoy, *Phys. Rev. A* **2**, 1075 (1970).
19. E. Barouch and B. M. McCoy, *Phys. Rev. A* **3**, 786 (1971).
20. E. Lieb, T. Schultz and D. Mattis, *Ann. Phys.* **16**, 407 (1961).
21. L. Amico *et al.*, *Rev. Mod. Phys.* **80**, 517 (2008).
22. G. Giorgi, *Phys. Rev. B* **79**, 060405(R) (2009).
23. A. De Pasquale and P. Facchi, *Phys. Rev. A* **80**, 032102 (2009).
24. M. Continentino, *Quantum Scaling in Many Body Systems* (World Scientific, Singapore, 2001).
25. F. Franchini *et al.*, *J. Phys. A: Math. Theor.* **40**, 8467 (2007).
26. G. Toth, private communication.
27. O. F. Syljuåsen, *Phys. Rev. A* **68**, 060301(R) (2003).
28. A. Osterloh *et al.*, *Phys. Rev. Lett.* **97**, 257201 (2006).
29. T. R. de Oliveira *et al.*, *Phys. Rev. A* **77**, 032325 (2008).
30. J. D. Johnson and B. M. McCoy, *Phys. Rev. A* **4**, 2314 (1971).
31. U. Schöllwock, *Rev. Mod. Phys.* **77**, 259 (2005).
32. R. Dillenschneider, *Phys. Rev. B* **78**, 224413 (2008).
33. M. S. Sarandy, *Phys. Rev. A* **80**, 022108 (2009).
34. S. Luo, *Phys. Rev. A* **77**, 042303 (2008).
35. J. Maziero *et al.*, *Phys. Rev. A* **82**, 012106 (2010).
36. L. Ciliberti, R. Rossignoli and N. Canosa, *Phys. Rev. A* **82**, 042316 (2010).
37. A. Hamma *et al.*, *Phys. Rev. A* **71**, 022315 (2005).
38. A. Kitaev and J. Preskill, *Phys. Rev. Lett.* **96**, 110404 (2006).
39. M. Levin and X.-G. Wen, *Phys. Rev. Lett.* **96**, 110405 (2006).
40. D. Abasto, A. Hamma and P. Zanardi, *Phys. Rev. A* **78**, 010301(R) (2008).
41. X. Chen, Z.-C. Gu and X.-G. Wen, *Phys. Rev. B* **82**, 155138 (2010).

Chapter 3

ENTANGLEMENT FROM THE DYNAMICS OF AN IDEAL BOSE GAS IN A LATTICE

SOUGATO BOSE

Department of Physics and Astronomy,
University College London, Gower St., London WC1E 6BT, UK

We show how the remotest sites of a finite lattice can be entangled, with the amount of entanglement exceeding that of a singlet, solely through the dynamics of an ideal Bose gas in a special initial state in the lattice. When additional occupation number measurements are made on the intermediate lattice sites, then the amount of entanglement and the length of the lattice separating the entangled sites can be significantly enhanced. The entanglement generated by this dynamical procedure is found to be higher than that for the ground state of an ideal Bose gas in the same lattice. A second dynamical evolution is shown to verify the existence of these entangled states, as well entangle qubits belonging to well separated quantum registers.

Keywords: Entanglement; bosons; optical lattice.

1. Introduction

One of the aims in the field of quantum information is to set up entanglement between locations separated by some distance, and in general, greater the separation and more the *amount* of this entanglement, the better. While photons are best for long distance entanglement distribution, for short distances (such as for linking quantum registers) other alternatives are important.[1] In this context, the *dynamics* of spin chains have been proposed for the distribution of entanglement over a distance of several lattice sites (For example, Refs. 2–8 to mention a very few). However, as the number of possible states of a spin in a spin chain is low, the amount of entanglement that can be dynamically generated and distributed through a single spin chain channel in a limited time is restricted. In this paper, as an alternative to spin chains, we suggest the use of the dynamics of an ideal gas of M bosons in a lattice to generate and distribute entanglement between its remotest sites. Note that our dynamics, when followed by certain occupation number measurements, will create high but "finite" dimensional entangled states (with entanglement $\sim \log_2 \sqrt{M} + \text{Const.}$) which are qualitatively very different from the infinite dimensional Gaussian entangled states which can be generated dynamically through harmonic oscillator chains.[9,10]

Another motivation for the current work originates from the literature on entangling Bose–Einstein condensates (BECs) of gaseous atoms/molecules in distinct traps (Refs. 11–16 to mention a very few) where usually small lattices or continuous variable entanglement are considered. Can we use lattice dynamics to create entanglement between traps separated by *several intervening lattice sites*? Here we show how to accomplish this without either the physical movement of traps or any local modulation of the lattice parameters.

We consider a one dimensional lattice of N sites, where the aim is to establish a significant amount of entanglement between sites 1 and N. We choose N to be odd and initially place M bosons in the $(N+1)/2$th site of the lattice and keep all the other sites empty. Physically this corresponds to a Fock state $|M\rangle$ on the $(N+1)/2$th site of the lattice and a vacuum state in all the other sites. In terms of boson creation operators $a_j^\dagger$ which create a boson in the jth lattice site the initial state is thus

$$|\Psi(0)\rangle = \frac{\left(a_{\frac{N+1}{2}}^\dagger\right)^M}{\sqrt{M!}}|0\rangle. \tag{1}$$

This special initial state greatly simplifies the calculations and can be regarded as a generalization of a single spin flip at the midpoint of a spin chain, which has been studied for entanglement generation.[8] We assume the bosons to be essentially noninteracting during the time-scale of our scheme i.e., their collection is an *ideal Bose gas*. Then the Hamiltonian of the system in the lattice is:

$$H = J \sum_{j=1}^{N} \left(a_j^\dagger a_{j+1} + a_j a_{j+1}^\dagger\right). \tag{2}$$

The state of each boson then evolves independently in the lattice (i.e., each boson evolves as if it was hopping alone in an otherwise empty lattice) as $a_{(N+1)/2}^\dagger \to \sum_{j=1}^{N} f_j(t)a_j^\dagger$ where $f_j(t)$ is the amplitude of the transfer of a single boson from the $(N+1)/2$th site to the jth site in time t. Thus the state of the M boson system at time t is:

$$|\Psi(t)\rangle = \frac{\left(\sum_{j=1}^{N} f_j(t)a_j^\dagger\right)^M}{\sqrt{M!}}|0\rangle. \tag{3}$$

The evolution amplitudes $f_j(t)$ are identical to that of a XY spin chain in the single excitation sector, and is given[7,17] by:

$$f_j(t) = \frac{2}{N+1} \sum_{k=1}^{N} \left\{\sin\frac{\pi k}{2} \sin\frac{\pi k j}{N+1}\right\} e^{i2Jt\cos\frac{k\pi}{N+1}}. \tag{4}$$

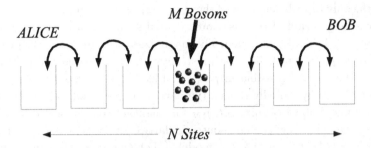

Fig. 1. Our setup of creating entanglement between sites 1 and N of a 1D lattice. One simply starts with M bosons in the $(N+1)/2$th lattice sites and allows dynamical evolution of the system to create entanglement between the sites 1 and N. Alice and Bob who have access to the sites 1 and N can use the entanglement for quantum communications for linking distinct quantum registers.

Equations (3) and (4) give the complete time evolution of the M boson state analytically.

Our task is now to calculate how much entanglement exists between sites 1 and N in the state of Eq. (3) as a function of time and find a time at which this entanglement is large. We will thus have to calculate the reduced density matrix of the states of sites 1 and N by tracing out the state of all the other sites. To accomplish that we adopt a strategy from Ref. 20, which evaluates the entanglement between two regions of a Bose–Einstein condensate in its ground state in a *single* trap. The strategy is to define new creation/annihilation operators by combining the operators $a_j^\dagger$ as $E^\dagger = f_1(t)(a_1^\dagger + a_N^\dagger)/\sqrt{2|f_1(t)|^2}$ and $L^\dagger = \sum_{j=2}^{N-1} f_j(t)a_j^\dagger/\sqrt{1-2|f_1(t)|^2}$, which are valid bosonic creation operators satisfying $[E, E^\dagger] = 1$, $[L, L^\dagger] = 1$. Noting that the symmetry of our problem implies $f_1(t) = f_N(t)$, we expand the expression of $|\Psi(t)\rangle$ in Eq. (3) in terms of the above operators to get:

$$|\Psi(t)\rangle = \sum_{r=0}^{M} \sqrt{^M C_r}(\sqrt{2}|f_1(t)|)^r(\sqrt{1-2|f_1(t)|^2})^{M-r} \times |\psi_r\rangle_{1N}|\phi_r\rangle_{2\cdots N}, \quad (5)$$

where we have substituted $|\psi_r\rangle_{1N} = ((E^\dagger)^r/\sqrt{r!})|0\rangle$ and $|\phi_r\rangle_{2\cdots N} = (L^\dagger)^{M-r}/\sqrt{(M-r)!}|0\rangle$. Noting that the set of states $\{|\phi_r\rangle_{2\cdot N}\}$ represents an orthonormal set in the space of states of the sites 2 to $N-1$, we have the reduced density matrix of sites 1 and N to be:

$$\rho(t)_{1N} = \sum_{r=0}^{M} P_r(t)|\psi_r\rangle\langle\psi_r|_{1N}, \quad (6)$$

where $P_r(t) = {}^M C_r(2|f_1(t)|^2)^r(1-2|f_1(t)|^2)^{M-r}$. We can write $|\psi_r\rangle_{1N}$ in terms of the occupation numbers as:

$$|\psi_r\rangle_{1N} = \frac{1}{2^{r/2}} \sum_{k=0}^{r} \sqrt{^r C_k}|k\rangle_1|r-k\rangle_N. \quad (7)$$

Note that the only time dependence of the state $\rho(t)_{1N}$ stems from $f_1(t)$, maximizing which over a long period of time is a pretty good strategy for obtaining a time t_h such that $\rho(t_h)_{1N}$ is highly entangled. The maximization ensures that the proportion of the state $|\psi_M\rangle_{1N}$, which has the most entanglement among the set of states $\{|\psi_r\rangle_{1N}\}$, is the highest possible in $\rho(t_h)_{1N}$. Ideally, the lattice dynamics should be frozen at t_h, say by globally raising the barriers between all wells of the lattice, so that Alice and Bob can utilize $\rho(t_h)_{1N}$ for quantum communications or linking quantum registers. For the smallest nontrivial lattice ($N = 3$) we know that $f_1(t) = 1/\sqrt{2}$ at $t = \pi/2J\sqrt{2}$ from the XY model.[17] For this case, the state $|\psi_M\rangle_{13}$ is generated between sites 1 and 3 at $t = \pi/2J\sqrt{2}$ whose entanglement can be made to grow without limits by increasing M. For this special case, the advantage over spin-1/2 chains is most evident, where only the case of $M = 1$ can be realized (with a single flip in the middle of a chain of three spin-1/2 systems).

In general (for $N > 3$), the state $\rho(t)_{1N}$ is a mixed state of a $(M+1) \times (M+1)$ dimensional system, and the only readily computable measure of its entanglement is the logarithmic negativity E_n,[18,19] which bounds the amount of pure state entanglement extractable by local actions from the state $\rho(t)_{1N}$. It is the standard measure used when high dimensional mixed entangled states arise.[9,10] The E_n of $\rho(t)_{1N}$ for an appropriately chosen time $t = t_h$ (see above) are plotted in Fig. 2 as a function of N for different values of M. This figure clearly shows that for N as high as 21, one can generate more entanglement than that of a singlet ($E_n = 1$ for a singlet). For such modest lengths, thus one generates more entanglement between the ends of a lattice by using a $M > 1$ boson gas than ever possible with a spin-1/2 chain, which is the $M = 1$ case, also plotted in Fig. 2. The advantages of increasing

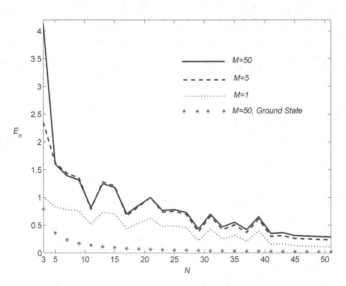

Fig. 2.   Entanglement (log negativity) of sites 1 and N from dynamics at an optimal time for various N and M, and for the ground state.

M diminish, though, as one increases N. For $N = 51$, we see that though there is some advantage of high boson number ($M = 50$) over the spin chain ($M = 1$) case, this advantage does not increase by increasing M (for example, the $M = 5$ and 50 plots are nearly coincident).

Next, we slightly modify our scheme and after the dynamical evolution till t_h, we measure the total number of bosons in the intermediate sites (sites 2 to $N - 1$). With probability P_r, we will find this number to be $M - r$, and when we do so, we will generate the state $|\psi_r\rangle_{1N}$ between sites 1 and N. Note that $|\psi_r\rangle_{1N}$ is created between sites 1 and N whenever a "total" of $M - r$ bosons is found in the remaining sites *irrespective* of the distribution of these $M - r$ bosons among the sites. The amount of entanglement in the pure state $|\psi_r\rangle_{1N}$ is given by its von Neumann entropy of entanglement $E_v(|\psi_r\rangle_{1N}) = -\mathrm{Tr}(\rho_1 \log_2 \rho_1)$, where $\rho_1 = \mathrm{Tr}_N(|\psi_r\rangle\langle\psi_r|_{1N})$, which *equals* the quantity of entanglement that can be obtained as singlets (the most useful form, say for their use as a resource for perfect teleportation of qubits from Alice to Bob) from the state by local actions and classical communications alone. Thus the average von Neumann entropy of entanglement $\langle E_v\rangle = \sum_r P_r E_v(|\psi_r\rangle_{1N})$ over all possible measurement outcomes is operationally the most useful measure of the entanglement between sites 1 and N in our modified scheme. This quantity has been plotted in Fig. 3 for various M for large lengths of lattice up to $N = 1001$. We find that for $M = 1000$, entanglement nearly equal to that of four singlets is generated across a distance of 1001 lattice sites, and this is more than 70 times the amount possible with a spin-1/2 chain ($M = 1$) of same length. For high N and M, we can represent P_r by a Poisson distribution and know that

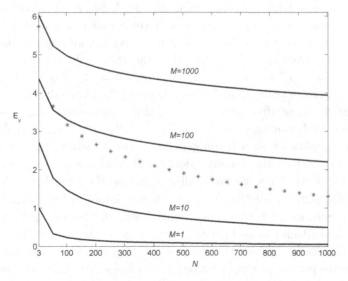

Fig. 3.   Average entanglement (von Neumann entropy) of sites 1 and N from dynamics followed by measurements on the intermediate lattice sites for various N and M. The plot with asterisks denotes the same for the ground state for $M = 1000$.

$f_1(t) \sim 1.7/N^{1/3}$ at $t \sim (N + 0.81N^{1/3})/4J$.[2-6,17] Using these, we have the analytic expression $\langle E_v \rangle \sim \log_2\{1.7\sqrt{M\pi e}/N^{1/3}\}$, whose fit with data gets better as we proceed from the $M = 10$ to the $M = 1000$ plot at high N. For every order of magnitude increase of M we thus expect to gain an entanglement equalling that of $\log_2 \sqrt{10} = 1.66$ singlets, which is confirmed by the differences in $\langle E_v \rangle$ between the $M = 100$ and 1000 plots at high N (Fig. 3). So for obtaining entanglement exceeding that of 10 singlets across a distance of ~ 1000 lattice sites we require $M \sim 10^7$.

Now we proceed to compare the amount of entanglement between sites 1 and N generated by the above schemes with that obtainable from the ground state of an ideal Bose gas in the same lattice (to check whether we have gained from dynamics). Ground state entanglement between distinct regions of a Bose–Einstein condensate in a *single* trap has already been investigated.[20,23] In our lattice setting, the ground state of H is simply $((1/\sqrt{N})\sum_{j=1}^{N} a_j^\dagger)^M |0\rangle/\sqrt{M!}$. It thus suffices to replace $f_1(t)$ in the expression of $P_r(t)$ with $1/\sqrt{N}$ to change from the dynamical state to the ground state. E_n between sites 1 and N for the ground state is plotted using asterisks in Fig. 2 as a function of N for $M = 50$. We find that the entanglement of sites 1 and N is nearly vanishing for $N \geq 25$, and thus there is a marked advantage of using dynamics as opposed to the ground state. One may also however, measure the occupation numbers of sites 2 to N of the ground state to obtain an average entanglement $\langle E_v \rangle \sim \log_2 \sqrt{M\pi e/N}$ between sites 1 and N (for high M and N). Even this is lower than that of our second scheme by $\log_2 \{1.7N^{1/6}\}$.

We now proceed to discuss a method to verify the pure entangled states $|\psi_r\rangle_{1N}$ created through our second scheme. One can easily verify the number correlations (the fact that if site 1 has k bosons, then the site N has $r - k$ bosons) in $|\psi_r\rangle_{1N}$ by occupation number measurements. Then the only task remaining is to verify the coherence between the terms $|k\rangle_1|r - k\rangle_N$ in $|\psi_r\rangle_{1N}$. If we apply a Hamiltonian $H_v = \kappa a_1^\dagger a_1$ for a fixed time t_v to site 1 then the above coherence leads to $|\psi_r\rangle_{1N}$ becoming $|\psi_r'\rangle_{1N} = (e^{i\theta}a_1^\dagger + a_N^\dagger)^r|0\rangle/\sqrt{2^r r!}$ where $\theta = \kappa t_v$. The ideal Bose gas is then allowed to evolve again in the lattice (assuming the lattice sites 2 to $N-1$ have been emptied due to or after the earlier occupation number measurements), which results in its state becoming $|\Psi'(t)\rangle_{1..N} = (\sum_{j=1}^{N}\{e^{i\theta}g_{1j}(t) + g_{Nj}(t)\}a_j^\dagger)^r|0\rangle/\sqrt{2^r r!}$ where g_{lj} are amplitudes for a boson going from the lth site to the jth site. From symmetry and the formulae for evolution in XY chain, we have $g_{1(N+1)/2}(t) = g_{N(N+1)/2}(t) = f_1(t)$. Thus the probability of finding the site $(N+1)/2$ occupied can be varied from $1 - (1 - 2|f_1(t)|^2)^r$ for $\theta = 0$, to 0 for $\theta = \pi$ (as the term $a_{(N+1)/2}^\dagger$ in $|\Psi'(t)\rangle_{1..N}$ vanishes for $\theta = \pi$). This variation with θ, whose range increases with increasing r and which can be further increased by maximizing $|f_1(t)|$, enables us to verify the coherence between the terms $|k\rangle_1|r - k\rangle_N$ in $|\psi_r\rangle_{1N}$. One way to verify the mixed state $\rho(t)_{1N}$ produced by our first scheme will be to go through our second scheme and check that pure states $|\psi_r\rangle_{1N}$ are produced with probabilities P_r.

Next, we provide an example of linking qubits A and B of distant quantum registers by making them interact with sites 1 and N respectively for a time t_q, when these sites are in the state $|\psi_r\rangle_{1N}$. The initial state of the qubits is $(|0\rangle_A + |1\rangle_A) \otimes (|0\rangle_B + |1\rangle_B)/2$ and the interaction Hamiltonian is $H_q = g(\sigma_z^A a_1^\dagger a_1 + \sigma_z^B a_N^\dagger a_N)$ in which σ_z^A and σ_z^B are Pauli operators of the qubits. For $gt_q = \pi$, the state of the system is $\{(|0\rangle_A|0\rangle_B + (-1)^r|1\rangle_A|1\rangle_B) \otimes (a_1^\dagger + a_N^\dagger)^r|0\rangle + (|0\rangle_A|1\rangle_B + (-1)^r|1\rangle_A|0\rangle_B) \otimes (a_1^\dagger - a_N^\dagger)^r|0\rangle\}/\sqrt{2^{r+2}r!}$. As in the previous paragraph, now the state of the ideal Bose gas is again allowed to evolve in the lattice and after some time the presence of any boson the $(N+1)/2$th site is measured. If this site is found occupied, then, remembering the logic of the last paragraph, the gas must have been in the state $(a_1^\dagger + a_N^\dagger)^r|0\rangle/\sqrt{2^r r!}$, which projects the qubits to the maximally entangled state $|0\rangle_A|0\rangle_B + (-1)^r|1\rangle_A|1\rangle_B$. The probability for this is $1/2$. Otherwise (also with probability $1/2$), a state whose fidelity with $|0\rangle_A|1\rangle_B + (-1)^r|1\rangle_A|0\rangle_B$ is $1 - (1 - 2|f_1(t)|^2)^r$ is obtained (this fidelity can be maximized by maximizing $|f_1(t)|$, and gets better with increasing r). This scheme for entangling qubits of distinct registers is just an example. The values of $\langle E_v \rangle$ found earlier imply that in *principle* one should be able to extract many more singlets from $|\psi_r\rangle_{1N}$ by local actions alone.

BECs of dilute atomic gases in optical lattices form a test ground for our protocols. Fock states can be prepared by Mott transitions,[24,25] the interactions needed for that can be switched off by a Feshbach resonance[26,27] to obtain an ideal Bose gas (otherwise, one simply has to go between regimes were the on-site repulsion $U \gg J/M^2$ to $U \ll J/M^2 N$ by the global modulation of lattice potentials). Accurate measurement of the total number of bosons in sites 2 to $N-1$ is possible either by using metastable atoms[28] or potentially through a second Mott transition involving these sites only (note that individual site occupation numbers are not required). For the verification part, only whether the $(N+1)/2$th site is occupied or not need to be ascertained, and atomic fluorescence in external fields can potentially be used. By placing all atoms in a known magnetic state when required (say by a laser), each site can be imparted a magnetic moment proportional to its occupation number. This, and a local magnetic field at site 1 can be used to realize H_v. This magnetic moment also enables magnetic moment based register qubits (atomic or solid state) to interact with sites 1 and N through H_q. If imparting the atoms with a magnetic moment automatically make $U \gg J/M^2$ then we have to ensure that Ut_v and Ut_q are integral multiples of 2π. An alternative implementation is nano-oscillators arrays[9,10] when resoncoupled to each other. They can be coupled to Cooper-pair box qubits to both create and measure Fock states,[29] and to Josephson qubits through a Jaynes-Cummings model[30] whose off resonant limit can implement H_q. Coupled cavities in photonic crystals, where Fock state preparation and measurements could be performed with dopant atoms is another potential implementation.[31–33]

Acknowledgment

I thank EPSRC for an Advanced Research Fellowship and for support through the grant GR/S62796/01 and the QIPIRC (GR/S82176/01).

References

1. D. Kielpinski, C. Monroe and D. J. Wineland, *Nature* **417**, 709 (2002).
2. S. Bose, *Phys. Rev. Lett.* **91**, 207901 (2003).
3. D. Burgarth and S. Bose, *Phys. Rev. A* **71**, 052315 (2005)
4. V. Giovannetti and D. Burgarth, *Phys. Rev. Lett.* **96**, 030501 (2006).
5. T. Boness, S. Bose and T. S. Monteiro, *Phys. Rev. Lett.* **96**, 187201 (2006).
6. J. Fitzsimons and J. Twamley, *Phys. Rev. Lett.* **97**, 090502 (2006).
7. M. Christandl *et al.*, *Phys. Rev. Lett.* **92**, 187902 (2004).
8. M.-H. Yung and S. Bose, *Phys. Rev. A* **71**, 032310 (2005).
9. J. Eisert *et al.*, *Phys. Rev. Lett.* **93**, 190402 (2004).
10. M. B. Plenio, J. Hartley and J. Eisert, *New J. Phys.* **6**, 36 (2004).
11. J. A. Dunningham and K. Burnett, *Phys. Rev. A* **61**, 065601 (2000).
12. J. A. Dunningham, K. Burnett and M. Edwards, *Phys. Rev. A* **64**, 015601 (2001).
13. J. A. Dunningham and K. Burnett, *Phys. Rev. A* **70**, 033601 (2004).
14. A. P. Hines, R. H. McKenzie and G. J. Milburn, *Phys. Rev. A* **67**, 013609 (2003).
15. B. Deb and G. S. Agarwal, *Phys. Rev. A* **67**, 023603 (2003).
16. J. A. Dunningham *et al.*, quant-ph/0608242.
17. M.-H. Yung, D. W. Leung and S. Bose, *Quantum Inf. Comput.* **4**, 174 (2003).
18. G. Vidal and R. F. Werner, *Phys. Rev. A* **65**, 032314 (2002).
19. M. B. Plenio, *Phys. Rev. Lett.* **95**, 090503 (2005).
20. C. Simon, *Phys. Rev. A* **66**, 052323 (2002).
21. L. Heaney, J. Anders and V. Vedral, quant-ph/0607069.
22. V. Vedral and D. Kaszilikowski, quant-ph/0606238.
23. D. Kaszilikowski *et al.*, quant-ph/0601089.
24. M. Greiner *et al.*, *Nature* **415**, 39 (2002).
25. C. Orzel *et al.*, *Science* **291**, 2386 (2001).
26. S. Inouye *et al.*, *Nature* **392**, 151 (1998).
27. J. L. Roberts *et al.*, *Phys. Rev. Lett.* **86**, 4211 (2001).
28. A. Robert *et al.*, *Science* **292**, 461 (2001).
29. E. K. Irish and K. Schwab, *Phys. Rev. B* **68**, 155311 (2003).
30. M. R. Geller and A. N. Cleland, *Phys. Rev. A* **71**, 032311 (2005).
31. D. G. Angelakis, M. F. Santos and S. Bose, quant-ph/0606159.
32. M. J. Hartmann, F. G. S. L. Brandão and M. B. Plenio, quant-ph/0606097.
33. A. D. Greentree *et al.*, cond-mat/0609050.

Chapter 4

ASPECTS OF THE RIEMANNIAN GEOMETRY
OF QUANTUM COMPUTATION

HOWARD E. BRANDT

U.S. Army Research Laboratory, Adelphi, MD
howard.e.brandt.civ@mail.mil

A review is given of some aspects of the Riemannian geometry of quantum computation in which the quantum evolution is represented in the tangent space manifold of the special unitary unimodular group $SU(2^n)$ for n qubits. The Riemannian right-invariant metric, connection, curvature, geodesic equation for minimal complexity quantum circuits, Jacobi equation and the lifted Jacobi equation for varying penalty parameter are reviewed. Sharpened tools for calculating the geodesic derivative are presented. The geodesic derivative may facilitate the numerical investigation of conjugate points and the global characteristics of geodesic paths in the group manifold, the determination of optimal quantum circuits for carrying out a quantum computation, and the determination of the complexity of particular quantum algorithms.

Keywords: Quantum computing; quantum circuits; quantum complexity; differential geometry; Riemannian geometry; geodesics; Lax equation; Jacobi fields; geodesic derivative.

1. Introduction

A quantum computation can be ideally represented by a unitary transformation acting on a quantum register consisting of a set of qubits representing the Hilbert space of the computational degrees of freedom of the quantum computer, and any unitary transformation can be faithfully represented by a network of universal quantum gates, such as two-qubit controlled-NOT gates and certain single-qubit gates. This is the basis of the quantum circuit model of quantum computation.[1] An important measure of the difficulty of performing a quantum computation is the number of quantum gates needed. A quantum algorithm is considered efficient if the number of required gates scales only polynomial (not exponentially) with the size of the problem. Few general principles are known for finding optimal quantum circuits for synthesizing unitary operations. Quantum circuit networks are usually analyzed using discrete methods, however potentially powerful continuous differential geometric methods are under development, using sub-Riemannian,[2-4] Riemannian,[5-8] also Finsler,[5] and sub-Finsler[9] geometries. Since unitary transformations are themselves continuous, this is perhaps not a surprising development. Using these differential geometric methods, optimal paths may be sought in Hilbert space for executing a quantum computation.

An innovative approach to the differential geometry of quantum computation and quantum circuit complexity analysis was introduced by Nielsen and collaborators.[5–8] A Riemannian metric was formulated on the special unitary unimodular group manifold of multi-qubit unitary transformations $SU(2^n)$, such that the metric distance between the identity and the desired unitary operator, representing the quantum computation, is equivalent to the number of quantum gates needed to represent that unitary operator, thereby providing a measure of the complexity associated with the corresponding quantum computation. The Riemannian metric was defined as a positive-definite bilinear form expressed in terms of the multi-qubit Hamiltonian. The analytic form of the metric was chosen to penalize all directions on the manifold not easily simulated by local gates. In this way, basic differential geometric concepts such as the Levi–Civita connection, geodesic path, Riemannian curvature, Jacobi fields and conjugate points can be associated with quantum computation. The unitary transformation expressing the quantum evolution is an exponential involving the Hamiltonian. The Hamiltonian can be expressed in terms of tensor products of the Pauli matrices which act on the qubits. The geodesic equation on the manifold follows from the connection and determines the local optimal Hamiltonian evolution corresponding to the unitary transformation representing the desired quantum computation. The optimal unitary evolution may follow by solving the geodesic equation. Useful upper and lower bounds on the associated quantum circuit complexity may be obtained.[5–8] The present work presents an expository review of some aspects of the Riemannian geometry of the special unitary unimodular group manifold associated with quantum computation. Such differential geometric approaches to quantum computation are currently preliminary and many details remain to be worked out.

2. Riemannian Geometry of $SU(2^n)$

A Riemannian metric can be chosen on the manifold of the Lie Group $SU(2^n)$ (special unitary group) of n-qubit unitary operators with unit determinant.[10–27] The traceless Hamiltonian is a tangent vector to a point on the group manifold of the n-qubit unitary transformation U. The Hamiltonian H is an element of the Lie algebra $su(2^n)$ of traceless $2^n \times 2^n$ Hermitian matrices[24–27] and is tangent to the evolutionary curve $e^{-iHt}U$ at time $t = 0$. (Here and throughout, units are chosen such that Planck's constant divided by 2π is $\hbar = 1$.)

The Riemannian metric (inner product) $\langle .,. \rangle$ is taken to be a right-invariant positive definite bilinear form $\langle H, J \rangle$ defined on tangent vectors (Hamiltonians, for example) H and J. Right invariance of the metric means that all right translations are isometries.[10–13,26,27] It follows that the Levi–Civita connection is also right invariant. Following Ref. 8, the n-qubit Hamiltonian H can be divided into two parts $P(H)$ and $Q(H)$, where $P(H)$ contains only one and two-body terms, and $Q(H)$ contains more than two-body terms. Terms corresponding to more than two-body interactions are to be penalized since they generally correspond to complex gates which are difficult to implement. Thus:

$$H = P(H) + Q(H),\tag{2.1}$$

in which P and Q are superoperators acting on H, and obey the following relations:

$$P + Q = I, \quad PQ = QP = 0, \quad P^2 = P, \quad Q^2 = Q,\tag{2.2}$$

where I is the identity. The Hamiltonian can be expressed in terms of tensor products of the Pauli matrices[1]:

$$\sigma_0 \equiv I \equiv \begin{bmatrix} 1 & 0 \\ 0 & 1 \end{bmatrix}, \quad \sigma_1 \equiv X \equiv \begin{bmatrix} 0 & 1 \\ 1 & 0 \end{bmatrix},$$

$$\sigma_2 \equiv Y \equiv \begin{bmatrix} 0 & -i \\ i & 0 \end{bmatrix}, \quad \sigma_3 \equiv Z \equiv \begin{bmatrix} 1 & 0 \\ 0 & -1 \end{bmatrix}.\tag{2.3}$$

They are Hermitian,

$$\sigma_i = \sigma_i^\dagger \quad i = 0, 1, 2, 3,\tag{2.4}$$

where $\dagger$ denotes the adjoint, and, except for σ_0, they are traceless,

$$\mathrm{Tr}\,\sigma_i = 0, \quad i \neq 0.\tag{2.5}$$

Their products are given by:

$$\sigma_i^2 = I.\tag{2.6}$$

Also,

$$\sigma_i \sigma_j = i\varepsilon_{ijk}\sigma_k, \quad i, j, k \neq 0,\tag{2.7}$$

expressed in terms of the totally antisymmetric Levi–Civita symbol with $\varepsilon_{123} = 1$, and using the Einstein sum convention. Quantum gates can be expressed in terms of tensor products of Pauli matrices.

For explicitness, consider the case of a 3-qubit Hamiltonian, for Pauli matrices, σ_1, σ_2, and σ_3,[8] one has:

$$\begin{aligned} P(H) = {} & x^1\sigma_1 \otimes I \otimes I + x^2\sigma_2 \otimes I \otimes I + x^3\sigma_3 \otimes I \otimes I \\ & + x^4 I \otimes \sigma_1 \otimes I + x^5 I \otimes \sigma_2 \otimes I + x^6 I \otimes \sigma_3 \otimes I \\ & + x^7 I \otimes I \otimes \sigma_1 + x^8 I \otimes I \otimes \sigma_2 + x^9 I \otimes I \otimes \sigma_3 \\ & + x^{10}\sigma_1 \otimes \sigma_2 \otimes I + x^{11}\sigma_1 \otimes I \otimes \sigma_2 + x^{12} I \otimes \sigma_1 \otimes \sigma_2 \\ & + x^{13}\sigma_2 \otimes \sigma_1 \otimes I + x^{14}\sigma_2 \otimes I \otimes \sigma_1 + x^{15} I \otimes \sigma_2 \otimes \sigma_1 \\ & + x^{16}\sigma_1 \otimes \sigma_3 \otimes I + x^{17}\sigma_1 \otimes I \otimes \sigma_3 + x^{18} I \otimes \sigma_1 \otimes \sigma_3 \\ & + x^{19}\sigma_3 \otimes \sigma_1 \otimes I + x^{20}\sigma_3 \otimes I \otimes \sigma_1 + x^{21} I \otimes \sigma_3 \otimes \sigma_1 \end{aligned}$$

$$+ x^{22}\sigma_2 \otimes \sigma_3 \otimes I + x^{23}\sigma_2 \otimes I \otimes \sigma_3 + x^{24}I \otimes \sigma_2 \otimes \sigma_3$$

$$+ x^{25}\sigma_3 \otimes \sigma_2 \otimes I + x^{26}\sigma_3 \otimes I \otimes \sigma_2 + x^{27}I \otimes \sigma_3 \otimes \sigma_2$$

$$+ x^{28}\sigma_1 \otimes \sigma_1 \otimes I + x^{29}\sigma_2 \otimes \sigma_2 \otimes I + x^{30}\sigma_3 \otimes \sigma_3 \otimes I$$

$$+ x^{31}\sigma_1 \otimes I \otimes \sigma_1 + x^{32}\sigma_2 \otimes I \otimes \sigma_2 + x^{33}\sigma_3 \otimes I \otimes \sigma_3$$

$$+ x^{34}I \otimes \sigma_1 \otimes \sigma_1 + x^{35}I \otimes \sigma_2 \otimes \sigma_2 + x^{36}I \otimes \sigma_3 \otimes \sigma_3, \qquad (2.8)$$

in which $\otimes$ denotes the tensor product,[1,28] and the i in x^i serves as an index,

$$Q(H) = x^{37}\sigma_1 \otimes \sigma_2 \otimes \sigma_3 + x^{38}\sigma_1 \otimes \sigma_3 \otimes \sigma_2$$

$$+ x^{39}\sigma_2 \otimes \sigma_1 \otimes \sigma_3 + x^{40}\sigma_2 \otimes \sigma_3 \otimes \sigma_1$$

$$+ x^{41}\sigma_3 \otimes \sigma_1 \otimes \sigma_2 + x^{42}\sigma_3 \otimes \sigma_2 \otimes \sigma_1$$

$$+ x^{43}\sigma_1 \otimes \sigma_1 \otimes \sigma_2 + x^{44}\sigma_1 \otimes \sigma_2 \otimes \sigma_1 + x^{45}\sigma_2 \otimes \sigma_1 \otimes \sigma_1$$

$$+ x^{46}\sigma_1 \otimes \sigma_1 \otimes \sigma_3 + x^{47}\sigma_1 \otimes \sigma_3 \otimes \sigma_1 + x^{48}\sigma_3 \otimes \sigma_1 \otimes \sigma_1$$

$$+ x^{49}\sigma_2 \otimes \sigma_2 \otimes \sigma_1 + x^{50}\sigma_2 \otimes \sigma_1 \otimes \sigma_2 + x^{51}\sigma_1 \otimes \sigma_2 \otimes \sigma_2$$

$$+ x^{52}\sigma_2 \otimes \sigma_2 \otimes \sigma_3 + x^{53}\sigma_2 \otimes \sigma_3 \otimes \sigma_2 + x^{54}\sigma_3 \otimes \sigma_2 \otimes \sigma_2$$

$$+ x^{55}\sigma_3 \otimes \sigma_3 \otimes \sigma_1 + x^{56}\sigma_3 \otimes \sigma_1 \otimes \sigma_3 + x^{57}\sigma_1 \otimes \sigma_3 \otimes \sigma_3$$

$$+ x^{58}\sigma_3 \otimes \sigma_3 \otimes \sigma_2 + x^{59}\sigma_3 \otimes \sigma_2 \otimes \sigma_3 + x^{60}\sigma_2 \otimes \sigma_3 \otimes \sigma_3$$

$$+ x^{61}\sigma_1 \otimes \sigma_1 \otimes \sigma_1 + x^{62}\sigma_2 \otimes \sigma_2 \otimes \sigma_2 + x^{63}\sigma_3 \otimes \sigma_3 \otimes \sigma_3. \qquad (2.9)$$

Here, all possible tensor products having one and two-qubit Pauli matrix operators on three qubits appear in $P(H)$, and analogously, all possible tensor products having three-qubit operators appear in $Q(H)$. Tensor products including only the identity are excluded because the Hamiltonian is taken to be traceless. Each of the terms in Eqs. (2.8) and (2.9) is an 8×8 matrix. The various tensor products of Pauli matrices such as those appearing in Eqs. (2.8) and (2.9) are referred to as generalized Pauli matrices. In the case of an n-qubit Hamiltonian, there are $4^n - 1$ possible traceless tensor products (corresponding to the dimension of the $SU(2^n)$ tangent space $T_U SU(2^n)$ and the $su(2^n)$ algebra), and each term is a $2^n \times 2^n$ matrix.

The right-invariant,[10–13,26,27] Riemannian metric for tangent vectors H and J is given by[8]

$$\langle H, J \rangle \equiv \frac{1}{2^n} \mathrm{Tr}[HP(J) + qHQ(J)]. \qquad (2.10)$$

Here q is a large penalty parameter (with $q > 2^n$) which taxes interactions involving more than two-body terms. Justification for the form of the metric, Eq. (2.10), is given in Refs. 5 and 8. The length of an evolutionary path on the $SU(2^n)$ manifold is given by the integral over time from some initial time to a final time and is a measure of the cost of applying a control Hamiltonian $H(t)$ along the path. The Riemannian distance between an initial point and a final point in the manifold is the infimum of the length of all curves connecting those points.[13] A geodesic curve is in general only locally minimizing.[13]

In order to obtain the Levi–Civita connection, it is necessary to exploit the Lie algebra $su(2^n)$ associated with the group $SU(2^n)$. Because of the right-invariance of the metric, if the Christoffel symbols are calculated at the origin, the same expression applies everywhere on the manifold. Following Ref. 8, consider the unitary transformation

$$U = e^{-iX} \tag{2.11}$$

in the neighborhood of the identity $I \in SU(2^n)$ (or equivalently in the neighborhood of the origin of the tangent space manifold) with

$$X = x \cdot \sigma \equiv \sum_\sigma x^\sigma \sigma, \tag{2.12}$$

which expresses symbolically terms like those in Eqs. (2.8) and (2.9) generalized to 2^n dimensions. In Eqs. (2.11) and (2.12), X is defined using the standard branch of the logarithm with a cut along the negative real axis. In Eq. (2.12), for the general case of n qubits, x^σ represents the set of real $(4^n - 1)$ coefficients of the generalized Pauli matrices σ which represent all of the n-fold tensor products. The generalized Pauli matrices serve as basis vectors in the tangent space manifold. It follows from Eq. (2.12) that the factor x^σ multiplying a particular generalized Pauli matrix σ is given by:

$$x^\sigma = \frac{1}{2^n} \text{Tr}(X\sigma). \tag{2.13}$$

These are so-called Pauli coordinates. (In the neighborhood of the origin, X will be represented as $X = \Delta x^\mu \mu$ for infinitesimal Δx^μ and generalized Pauli matrix μ, where the Einstein sum convention summing over μ is to be understood. [See Eq. (2.42)]

Consider a curve $e^{-iHt}e^{-iX}$ in the $SU(2^n)$ group manifold, evolving from a point $U = e^{-iX}$, and representing a system with initial action X acted on by a control Hamiltonian H. For the point U on the $SU(2^n)$ manifold with tangent vector H to the curve, one has in the neighborhood of the identity,

$$e^{-iHt}e^{-iX} = e^{-i(X+Jt)} + 0(t^2) \tag{2.14}$$

to second-order in the time t. This follows from the Baker–Campbell–Hausdorff formula.[16,22,29,30] The right side of Eq. (2.14) contains the resulting total action $(X + Jt)$. Explicitly, the matrix J, the so-called Pauli representation of the tangent vector in the Pauli-coordinate representation in the tangent space $T_U SU(2^n)$, is related to H, the Hamiltonian representation of the tangent vector, by:

$$H = E_X(J), \tag{2.15}$$

in which the linear superoperator E_X is given by:

$$E_X = i \text{ad}_X^{-1}(e^{-i\text{ad}_X} - I), \tag{2.16}$$

where I is the identity, a power series expansion is to be understood since the operator ad_X is not invertible, and $\mathrm{ad}_X(Y)$ is the Lie bracket, defined by the ordinary matrix commutator

$$\mathrm{ad}_X(Y) \equiv [X, Y]. \tag{2.17}$$

The power series expansion of E_X is:

$$E_X = \sum_{j=0}^{\infty} \frac{(-i\mathrm{ad}_X)^j}{(j+1)}. \tag{2.18}$$

Near the origin, E_X is invertible, and one has:

$$J \equiv D_X(H) = E_X^{-1}(H). \tag{2.19}$$

It then follows from Eqs. (2.18) and (2.19) near the origin that:

$$E_X = I - \frac{i}{2}\mathrm{ad}_X + O(X^2) \tag{2.20}$$

and

$$D_X = I + \frac{i}{2}\mathrm{ad}_X + O(X^2). \tag{2.21}$$

One also has the adjoint relations with respect to the trace inner product[8,31,32]:

$$E_X^\dagger = E_{-X}, \tag{2.22}$$

$$D_X^\dagger = D_{-X}. \tag{2.23}$$

Next, the right-invariant metric, Eq. (2.10), in the so-called Hamiltonian representation can also be written as:

$$\langle H, J \rangle = \frac{1}{2^n}\mathrm{Tr}(HG(J)), \tag{2.24}$$

in which the positive self-adjoint superoperator G is given by:

$$G = P + qQ. \tag{2.25}$$

It is also useful in the following to define a Hermitian matrix L, dual to the Hamiltonian H,

$$L = G(H), \tag{2.26}$$

so that Eq. (2.24) can also be written as:

$$\langle H, J \rangle \equiv \frac{1}{2^n}\mathrm{Tr}(LJ). \tag{2.27}$$

Now consider the metric $\langle Y, Z \rangle$ for tangent vector fields Y and Z in the neighborhood of the origin at point $U = e^{-iX}$ [See Eq. (2.42)]. This will enable computing derivatives of the metric for the purpose of calculating the Christoffel symbols. By Eq. (2.15), the so-called Hamiltonian representations $\{Y^H, Z^H\}$ of the vector fields are related to their so-called Pauli representations $\{Y^P, Z^P\}$ by:

$$Y^H = E_X(Y^P), \quad Z^H = E_X(Z^P). \tag{2.28}$$

Substituting Eqs. (2.28) in Eq. (2.27), one obtains:

$$\langle Y, Z \rangle = \frac{1}{2^n} \mathrm{Tr}(Y^H G(Z^H)) = \frac{1}{2^n} \mathrm{Tr}(E_X(Y^P) G \circ E_X(Z^P)), \tag{2.29}$$

or

$$\langle Y, Z \rangle = \frac{1}{2^n} \mathrm{Tr}(Y^P E_X^\dagger \circ G \circ E_X(Z^P)). \tag{2.30}$$

Equivalently,

$$\langle Y, Z \rangle = \frac{1}{2^n} \mathrm{Tr}(Y^P G_X(Z^P)), \tag{2.31}$$

where

$$G_X \equiv E_X^\dagger \circ G \circ E_X. \tag{2.32}$$

The metric can be rewritten in the familiar Riemannian tensor form $g_{\sigma\tau}$, in a coordinate basis, as follows. The vectors Y^P and Z^P in the Pauli representation can be written in terms of generalized Pauli-matrix basis vectors as:

$$Y^P = \sum_\sigma y^\sigma \sigma, \quad Z^P = \sum_\sigma z^\sigma \sigma \tag{2.33}$$

with Pauli coordinates y^σ and z^σ. Here σ, as an index, is used to refer to a particular tensor product appearing in the generalized Pauli matrix σ. This index notation, used throughout, is a convenient abbreviation for the actual numerical indices (e.g., in Eq. (2.8), the number 23 appearing in x^{23}, the coefficient of $\sigma_2 \otimes I \otimes \sigma_3$). Then substituting Eqs. (2.33) in Eq. (2.31), one obtains:

$$\langle Y, Z \rangle = \frac{1}{2^n} \mathrm{Tr} \left(\sum_\sigma y^\sigma \sigma G_X \left(\sum_\tau z^\tau \tau \right) \right), \tag{2.34}$$

or

$$\langle Y, Z \rangle = \sum_{\sigma\tau} g_{\sigma\tau} y^\sigma z^\tau, \tag{2.35}$$

in which the Pauli-coordinate representation of the metric tensor $g_{\sigma\tau}$ in the neighborhood of the origin is given by:

$$g_{\sigma\tau} = \frac{1}{2^n} \mathrm{Tr}(\sigma G_X(\tau)). \tag{2.36}$$

One has in the neighborhood of the origin:

$$G_X = \left(1 + \frac{i}{2} \mathrm{ad}_X \right) \circ G \circ \left(1 - \frac{i}{2} \mathrm{ad}_X \right) + 0(X^2), \tag{2.37}$$

or equivalently,

$$G_X = G + \frac{i}{2} [\mathrm{ad}_X, G] + 0(X^2). \tag{2.38}$$

Next one has for the partial derivative of $g_{\sigma\tau}$ with respect to x^μ:

$$g_{\sigma\tau,\mu} = \lim_{\Delta x \to 0} \frac{g_{\sigma\tau}(x + \Delta x^\mu) - g_{\sigma\tau}(x)}{\Delta x^\mu}, \tag{2.39}$$

where the comma followed by μ denotes the partial derivative $\partial/\partial x^\mu$. Using Eqs. (2.36) and (2.38), then Eq. (2.39) becomes:

$$g_{\sigma\tau,\mu} = \lim_{\Delta x \to 0} \frac{1}{2^n} \text{Tr} \frac{\sigma\left(G + \frac{i}{2}[\text{ad}_X, G]\right)(\tau) - \sigma G(\tau)}{\Delta x^\mu}, \tag{2.40}$$

or

$$g_{\sigma\tau,\mu} = \lim_{\Delta x \to 0} \frac{i}{2^{n+1}} \text{Tr} \frac{\sigma[X, G(\tau)] - \sigma[G(\tau), X]}{\Delta x^\mu}. \tag{2.41}$$

In the neighborhood of the origin, one has for infinitesimals Δx^μ, using the Einstein sum convention for repeated upper and lower indices,

$$X = \Delta x^\mu \mu, \tag{2.42}$$

and when the μ-component is substituted in Eq. (2.41), one obtains:

$$g_{\sigma\tau,\mu} = \frac{i}{2^{n+1}} \text{Tr}(\sigma[\mu, G(\tau)] - \sigma[G(\tau), \mu]). \tag{2.43}$$

Next expanding the commutators, and using the cyclic property of the trace, one obtains:

$$g_{\sigma\tau,\mu} = \frac{i}{2^{n+1}} \text{Tr}(2(G(\tau)\sigma\mu - \sigma G(\tau)\mu)), \tag{2.44}$$

or equivalently,

$$g_{\sigma\tau,\mu} = \frac{i}{2^{n+1}} \text{Tr}(2[G(\tau), \sigma]\mu). \tag{2.45}$$

However because any Riemannian metric tensor is symmetric, one has:

$$g_{\sigma\tau,\mu} = \frac{1}{2}(g_{\sigma\tau,\mu} + g_{\tau\sigma,\mu}) \tag{2.46}$$

and substituting Eq. (2.45) in Eq. (2.46), one obtains[8,31,32]:

$$g_{\sigma\tau,\mu} = \frac{i}{2^{n+1}} \text{Tr}\{([G(\sigma), \tau] + [G(\tau), \sigma])\mu\}. \tag{2.47}$$

The familiar form of the Levi–Civita connection of Riemannian geometry, in a coordinate basis, is given by the Christoffel symbols of the first kind, namely,[13,19]

$$\Gamma_{\mu\sigma\tau} = \frac{1}{2}(g_{\mu\sigma,\tau} + g_{\mu\tau,\sigma} - g_{\sigma\tau,\mu}). \tag{2.48}$$

Substituting Eq. (2.47) in Eq. (2.48), one obtains:

$$\Gamma_{\mu\sigma\tau} = \frac{1}{2}\frac{i}{2^{n+1}} \text{Tr}(([G(\mu), \sigma] + [G(\sigma), \mu])\tau + ([G(\mu), \tau] + [G(\tau), \mu])\sigma$$
$$- ([G(\sigma), \tau] + [G(\tau), \sigma])\mu) \tag{2.49}$$

and expanding the commutators, using the cyclic property of the trace and simplifying, this becomes:

$$\Gamma_{\mu\sigma\tau} = \frac{i}{2^{n+1}} \text{Tr}((\tau G(\sigma) - G(\sigma)\tau)\mu + (\sigma G(\tau) - G(\tau)\sigma)\mu), \tag{2.50}$$

or

$$\Gamma_{\mu\sigma\tau} = \frac{i}{2^{n+1}} \operatorname{Tr}(([\sigma, G(\tau)] + [\tau, G(\sigma)])\mu), \tag{2.51}$$

and again using the cyclic property of the trace, one obtains:

$$\Gamma_{\mu\sigma\tau} = \frac{i}{2^{n+1}} \operatorname{Tr}(\mu([\sigma, G(\tau)] + [\tau, G(\sigma)])). \tag{2.52}$$

The inverse metric is given by[8,31,32]:

$$g^{\sigma\tau} = \frac{1}{2^n} \operatorname{Tr}(\sigma F(\tau)). \tag{2.53}$$

It then follows that the Christoffel symbols of the second kind are given by[8,31,32]:

$$\Gamma^{\rho}_{\sigma\tau} = \frac{i}{2^{n+1}} \operatorname{Tr}(F(\rho)([\sigma, G(\tau)] + [\tau, G(\sigma)])) \tag{2.54}$$

in which one defines the inverse superoperator

$$F(\rho) \equiv G^{-1}(\rho). \tag{2.55}$$

Next, for a generic Riemannian connection Γ^j_{kl} and vectors Z and Y, written in a coordinate basis, one has the familiar equation for the covariant derivative of Z along Y:

$$(\nabla_Y Z)^j = \frac{\partial z^j}{\partial x^k} y^k + \Gamma^j_{kl} y^k z^l \tag{2.56}$$

in which the Einstein convention of summing over repeated indices is implicit. Replacing indices (j, k, l) by (σ, τ, λ), multiplying both sides of Eq. (2.56) by σ, and summing over σ yields:

$$\sum_{\sigma} \sigma(\nabla_Y Z)^{\sigma} = \sum_{\sigma\tau} y^{\tau} \sigma \frac{\partial z^{\sigma}}{\partial x^{\tau}} + \sum_{\sigma\tau\lambda} \sigma \Gamma^{\sigma}_{\tau\lambda} y^{\tau} z^{\lambda} \tag{2.57}$$

and substituting Eqs. (2.33) and (2.54) in Eq. (2.57), one obtains:

$$(\nabla_Y Z)^P \equiv \sum_{\tau} y^{\tau} \frac{\partial Z^P}{\partial x^{\tau}} + \sum_{\sigma\tau\lambda} \sigma \frac{i}{2^{n+1}} \operatorname{Tr}\{F(\sigma)([\tau, G(\lambda)] + [\lambda, G(\tau)])\} y^{\tau} z^{\lambda}. \tag{2.58}$$

The following identity is true[31]:

$$\sum_{\sigma} \sigma \operatorname{Tr}\{F(\sigma)[\tau, G(\lambda)]\} = 2^n F([\tau, G(\lambda)]), \tag{2.59}$$

so that Eq. (2.58) becomes:

$$(\nabla_Y Z)^P \equiv \sum_{\tau} y^{\tau} \frac{\partial Z^P}{\partial x^{\tau}} + \sum_{\tau\lambda} \frac{i}{2^{n+1}} 2^n (F([\tau, G(\lambda)]) + F([\lambda, G(\tau)])) y^{\tau} z^{\lambda}. \tag{2.60}$$

Then substituting Eqs. (2.33) in Eq. (2.60), and using the Einstein sum convention, one obtains the Pauli representation of the connection evaluated at the origin with the vector fields given in the Pauli representation, namely,[8]:

$$(\nabla_Y Z)^P \equiv y^{\tau} \frac{\partial Z^P}{\partial x^{\tau}} + \frac{i}{2}(F([Y^P, G(Z^P)]) + F([Z^P, G(Y^P)])). \tag{2.61}$$

To obtain the Hamiltonian representation of the connection, one has according to Eqs. (2.19) and (2.21) near the origin,

$$Z^P = D_X(Z^H) = \left(1 + \frac{i}{2}\mathrm{ad}_X\right)(Z^H). \tag{2.62}$$

Also, clearly,

$$\frac{\partial Z^P}{\partial r^\sigma} = \lim_{\Delta r \to 0} \frac{Z^P(x + \Delta x) - Z^P(x)}{\Delta r^\sigma}, \tag{2.63}$$

and substituting Eq. (2.62) in Eq. (2.63), then,

$$\frac{\partial Z^P}{\partial x^\sigma} = \lim_{\Delta x \to 0} \frac{\left(1 + \frac{i}{2}\mathrm{ad}_X\right)(Z^H(x + \Delta x)) - Z^H}{\Delta x^\sigma}, \tag{2.64}$$

or substituting Eqs. (2.17) and (2.42), and dropping the Einstein sum convention here only, then,

$$\frac{\partial Z^P}{\partial x^\sigma} = \lim_{\Delta x \to 0} \frac{\frac{i}{2}[\Delta x^\sigma \sigma, Z^H] + \frac{\partial Z^H}{\partial x^\sigma}\Delta x^\sigma}{\Delta x^\sigma} = \frac{i}{2}[\sigma, Z^H] + \frac{\partial Z^H}{\partial x^\sigma}. \tag{2.65}$$

Thus

$$Z^P_{,\sigma} = \frac{i}{2}[\sigma, Z^H] + Z^H_{,\sigma}, \tag{2.66}$$

or multiplying by y^σ, using Eq. (2.33), and restoring the Einstein sum convention, one has:

$$y^\sigma Z^P_{,\sigma} = y^\sigma Z^H_{,\sigma} + \frac{i}{2}[Y^P, Z^H]. \tag{2.67}$$

Next substituting Eq. (2.67) in Eq. (2.61), one obtains:

$$(\nabla_Y Z)^P = y^\sigma Z^H_{,\sigma} + \frac{i}{2}[Y^P, Z^H] + \frac{i}{2}F([Y^P, G(Z^P)] + [Z^P, G(Y^P)]). \tag{2.68}$$

But at the origin, it is true that:

$$(\nabla_Y Z)^H = (\nabla_Y Z)^P, \quad Y^H = Y^P, \quad Z^H = Z^P \tag{2.69}$$

and the components y^σ of Y are the same in both representations. Therefore using Eqs. (2.68) and (2.69), one obtains the Hamiltonian representation of the connection at the origin[8]:

$$(\nabla_Y Z)^H = y^\sigma Z^H_{,\sigma} + \frac{i}{2}([Y^H, Z^H] + F([Y^H, G(Z^H)] + [Z^H, G(Y^H)])), \tag{2.70}$$

in which $Z^H_{,\sigma} \equiv (\partial Z^H / \partial x^\sigma)$. Equation (2.70) gives the covariant derivative of the vector Z^H along the vector Y^H.

Next, in order to determine the geodesic equation, consider a curve passing through the origin with tangent vector Y^H having components $y^\sigma = dx^\sigma/dt$. According to Eq. (2.70) and the chain rule, the covariant derivative along the curve in the Hamiltonian representation is given by:

$$(D_t Z)^H \equiv (\nabla_Y Z)^H$$

$$= \frac{dZ^H}{dt} + \frac{i}{2}([Y^H, Z^H] + F([Y^H, G(Z^H)] + [Z^H, G(Y^H)])). \quad (2.71)$$

Because of the right-invariance of the metric, Eq. (2.71) is true on the entire manifold. Furthermore, for a right-invariant vector field Z^H, one has

$$\frac{dZ^H}{dt} = 0 \quad (2.72)$$

and substituting Eq. (2.72) in Eq. (2.71), one obtains:

$$(\nabla_Y Z)^H = \frac{i}{2}\{[Y^H, Z^H] + F([Y^H, G(Z^H)] + [Z^H, G(Y^H)])\}, \quad (2.73)$$

which is also true everywhere on the manifold.

A geodesic in $SU(2^n)$ is a curve $U(t)$ with tangent vector $H(t)$ parallel transported along the curve, to which it is also tangent, namely,

$$D_t H = 0. \quad (2.74)$$

However, according to Eq. (2.71) with $Y^H = Z^H = H$, one has:

$$D_t H = \frac{dH}{dt} + \frac{i}{2}([H, H] + F([H, G(H)] + [H, G(H)])), \quad (2.75)$$

which when substituting Eq. (2.74) becomes[8]:

$$\frac{dH}{dt} = -iF([H, G(H)]). \quad (2.76)$$

One can rewrite Eq. (2.76) using Eqs. (2.26) and (2.55),

$$L \equiv G(H) = F^{-1}(H) \quad (2.77)$$

and then noting that:

$$\frac{dL}{dt} = \frac{d}{dt}(F^{-1}(H)) = F^{-1}\left(\frac{dH}{dt}\right). \quad (2.78)$$

Thus substituting Eq. (2.76) in Eq. (2.78), one obtains:

$$\frac{dL}{dt} = -iF^{-1}(F([H, G(H)])), \quad (2.79)$$

or

$$\frac{dL}{dt} = -i[H, G(H)] \quad (2.80)$$

and again using Eq. (2.77), Eq. (2.80) becomes:

$$\frac{dL}{dt} = -i[H, L] = i[L, H]. \quad (2.81)$$

Furthermore, again using Eq. (2.77) in Eq. (2.81), one obtains the sought geodesic equation[8]:

$$\frac{dL}{dt} = i[L, F(L)].\qquad(2.82)$$

Equation (2.82) is a Lax equation, a well-known nonlinear differential matrix equation, and L and $iF(L)$ are Lax pairs.[34–40] It is well to emphasize that L is a $2^n \times 2^n$ matrix in the case of an n-qubit quantum register. Analytical solutions to Eq. (2.82) are given in Refs. 8 and 31 for the case of three qubits and for the case in which only an initial value is given without requiring a specific target unitary transformation. The two-point boundary value problem is addressed below.

From the connection, the Riemann curvature and sectional curvature can also be calculated.[31,32,41] A nonvanishing term in the sectional curvature was erroneously omitted in. Ref. 8 the corrected sectional curvature is given by:

$$K(X,Y) = -\frac{3}{4}\langle i[X,Y], i[X,Y]\rangle + \frac{1}{4}\langle B(X,Y) + B(Y,X), B(X,Y) + B(Y,X)\rangle$$

$$+ \frac{1}{2}\langle i[X,Y], B(X,Y) - B(Y,X)\rangle - \langle B(X,X), B(Y,Y)\rangle,\qquad(2.83)$$

where

$$B(X,Y) = F(i[G(X), Y]),\qquad(2.84)$$

Equation (2.83) corrects Eq. (A.7) in Ref. 8.

Jacobi fields can be useful in the analysis of geodesics. For this purpose, consider a base geodesic $U \equiv \gamma^\sigma \sigma$ with coordinates $\gamma^\sigma(q,t)$ on the $SU(2^n)$ group manifold with penalty parameter q, and a neighboring geodesic with coordinates $\gamma^\sigma(q+\Delta, t)$ and penalty parameter $q + \Delta$. To first order in Δ, one has:

$$\gamma^\sigma(q + \Delta, t) = \gamma^\sigma(q, t) + \Delta J^\sigma(t),\qquad(2.85)$$

in which the Jacobi field $J = J^\sigma \sigma$ has coordinates $J^\sigma(t)$ defined by:

$$J^\sigma(t) = \frac{\partial \gamma^\sigma(q,t)}{\partial q}.\qquad(2.86)$$

The Hamiltonian for a geodesic $U \equiv \gamma(t)$ with penalty parameter q is given by:

$$H_q = \frac{d\gamma}{dt}\qquad(2.87)$$

for $t = 0$. Then taking the derivative with respect to the penalty parameter q, one has:

$$\frac{dH_q}{dq} = \frac{d}{dq}\frac{d\gamma}{dt},\qquad(2.88)$$

or equivalently,

$$\frac{dH_q}{dq} = \frac{d}{dt}\frac{d\gamma}{dq}\qquad(2.89)$$

and substituting Eq. (2.86) in Eq. (2.89), then for $t = 0$ one has:

$$\frac{dH_q}{dq} = \frac{dJ}{dt},$$

(2.90)

in which the Jacobi field J is:

$$J = J^\sigma \sigma.$$

(2.91)

To proceed, a lifted Jacobi equation can be obtained by varying the geodesic equation with respect to the penalty parameter q. The geodesic equation for the base geodesic with penalty parameter q is given by Eq. (2.81), namely,

$$\frac{dL}{dt} = i[L, H],$$

(2.92)

where the dual L is given by Eq. (2.26), namely,

$$L = G(H).$$

(2.93)

The geodesic equation for a nearby geodesic with penalty parameter $q + \Delta$ for small Δ is given by:

$$\frac{d\bar{L}}{dt} = i[\bar{L}, \bar{H}],$$

(2.94)

where

$$\bar{L} = \bar{G}(\bar{H}),$$

(2.95)

and in accord with Eq. (2.25), the superoperator G becomes:

$$\bar{G} = P + (q + \Delta)Q = G + \Delta\frac{dG}{dq} \equiv G + \Delta G',$$

(2.96)

in which one defines

$$G' \equiv \frac{dG}{dq} = Q,$$

(2.97)

according to Eq. (2.25).

Next letting $U(t)$ and $\bar{U}(t)$ denote the geodesics for penalty parameter q and $q + \Delta$, respectively, then for small Δ one expects[8]:

$$\bar{U} = Ue^{-i\Delta J},$$

(2.98)

and it follows that to first order in Δ,

$$\begin{aligned}
\frac{d\bar{U}}{dt} &= \frac{dU}{dt}e^{-i\Delta J} + U\frac{d}{dt}(1 - i\Delta J + 0(\Delta^2)) \\
&= \frac{dU}{dt}e^{-i\Delta J} + U\left(-i\Delta\frac{dJ}{dt} + 0(\Delta^2)\right) \\
&= \frac{dU}{dt}e^{-i\Delta J} - iU\Delta\frac{dJ}{dt}.
\end{aligned}$$

(2.99)

But according to the Schrödinger equation, one has:

$$\frac{d\bar{U}}{dt} = -i\overline{HU} \tag{2.100}$$

and

$$\frac{dU}{dt} = -iHU\,, \tag{2.101}$$

so substituting Eqs. (2.100) and (2.101) in Eq. (2.99), one obtains:

$$-i\overline{HU} = -iHUe^{-i\Delta J} - iU\Delta\frac{dJ}{dt}\,, \tag{2.102}$$

or substituting Eq. (2.98), then Eq. (2.102) for small Δ becomes:

$$-i\bar{H}Ue^{-i\Delta J} = -iHUe^{-i\Delta J} - iU\Delta\frac{dJ}{dt}\,, \tag{2.103}$$

or equivalently, then to order Δ,

$$-i\bar{H}U = -iHU - iU\Delta\frac{dJ}{dt}e^{i\Delta J} = -iHU - iU\Delta\frac{dJ}{dt}\,. \tag{2.104}$$

Next multiplying Eq. (2.104) on the right by $U^\dagger$ and noting that unitarity requires

$$UU^\dagger = 1\,, \tag{2.105}$$

then Eq. (2.104) becomes:

$$\bar{H} = H + \Delta U\frac{dJ}{dt}U^\dagger \tag{2.106}$$

to first order in Δ. Next substituting Eq. (2.95) in Eq. (2.94), one obtains:

$$\frac{d}{dt}(\bar{G}(\bar{H})) = i[\bar{G}(\bar{H})\,,\bar{H}]. \tag{2.107}$$

Then substituting Eqs. (2.96) and (2.106) in the left side of Eq. (2.107), the left side becomes:

$$\frac{d}{dt}(\bar{G}(\bar{H})) = \frac{d}{dt}((G + \Delta G')(H + \Delta K))\,, \tag{2.108}$$

where it is useful to define:

$$K \equiv U\frac{dJ}{dt}U^\dagger\,. \tag{2.109}$$

The lifted Jacobi equation will first be expressed in terms of K, and Eq. (2.109) will be used to obtain the lifted Jacobi field J in terms of K. To proceed then, Eq. (2.108) to first order in Δ becomes:

$$\frac{d}{dt}(\bar{G}(\bar{H})) = \frac{d}{dt}G(H) + \Delta\frac{d}{dt}(G'(H) + G(K)). \tag{2.110}$$

Next, using Eqs. (2.96), (2.106) and (2.109), the right side of Eq. (2.107) becomes:

$$i[\bar{G}(\bar{H}),\bar{H}] = i[(G + \Delta G')(H + \Delta K), H + \Delta K]\,, \tag{2.111}$$

or equivalently to first order in Δ,

$$i[\bar{G}(\bar{H}), \bar{H}] = i[G(H), H] + \Delta(i[G(H), K] + i[G'(H), H] + i[G(K), H]) . \qquad (2.112)$$

In terms of the dual, Eq. (2.93), Eq. (2.112) becomes:

$$i[\bar{G}(\bar{H}), \bar{H}] = i[L, H] + \Delta(i[L, K] + i[G'(H), H] + i[G(K), H]) . \qquad (2.113)$$

Next substituting Eqs. (2.108), (2.93) and (2.113) in Eq. (2.107), one obtains:

$$\frac{d}{dt}L + \Delta\left(G'\left(\frac{d}{dt}H\right) + G\left(\frac{d}{dt}K\right)\right)$$
$$= i[L, H] + \Delta(i[L, K] + i[G'(H), H] + i[G(K), H]) , \qquad (2.114)$$

and further substituting Eq. (2.92) in Eq. (2.114), one concludes that:

$$G'\left(\frac{d}{dt}H\right) + G\left(\frac{d}{dt}K\right) = i[L, K] + i[G'(H), H] + i[G(K), H] . \qquad (2.115)$$

Furthermore, multiplying Eq. (2.115) on the left by G^{-1}, one obtains:

$$G^{-1}G'\left(\frac{d}{dt}H\right) + \frac{d}{dt}K = G^{-1}(i[L, K] + i[G'(H), H] + i[G(K), H]) . \qquad (2.116)$$

But according to Eqs. (2.92) and (2.93), the geodesic equation yields:

$$\frac{d}{dt}H = iG^{-1}[L, H] = iF([L, H]) , \qquad (2.117)$$

so that, using Eqs. (2.26) and (2.55), one has:

$$G^{-1}G'\left(\frac{d}{dt}H\right) = FG'\left(\frac{d}{dt}H\right) = iFG'F([L, H]) . \qquad (2.118)$$

Then substituting Eqs. (2.118) and (2.26) in Eq. (2.116), one obtains:

$$0 = iFG'F([L, H]) + \frac{d}{dt}K + F(i[K, L] + i[H, G'(H)] + i[H, G(K)]) , \qquad (2.119)$$

or rearranging terms, then

$$0 = \frac{d}{dt}K + F(i[K, L] + i[H, G(K)] + G'F(i[L, H]) + i[H, G'(H)]) . \qquad (2.120)$$

Equation (2.120) is the lifted Jacobi equation for penalty parameter varied from q to $q + \Delta$.[8] It is an inhomogeneous first order nonlinear differential matrix equation in K. According to Eq. (2.97), if G is independent of q, then $G' = 0$, and the inhomogeneous terms vanishes. Then Eq. (2.120) reduces effectively to the conventional Jacobi equation, assuming the form,

$$0 = \frac{d}{dt}K + F(i[K, L] + i[H, G(K)]) . \qquad (2.121)$$

The inhomogeneous term in Eq. (2.120) is given by:

$$C = F(G'F(i[L, H]) + i[H, G'(H)]) . \qquad (2.122)$$

Substituting Eqs. (2.97), (2.26), (2.25) and (2.2) in Eq. (2.122), one has:

$$\begin{aligned}
C &= FQFi[(P+qQ)(H), P(H)+Q(H)] + Fi[H, Q(H)] \\
&= FQFi([P(H), Q(H)] + q[Q(H), P(H)]) + Fi([P(H)+Q(H), Q(H)]) \\
&= FQFi(1-q)[P(H), Q(H)] + Fi[P(H), Q(H)].
\end{aligned}$$

(2.123)

Equation (2.123) can be rewritten as follows:

$$\begin{aligned}
C &= F(P \mid Q)Fi(1-q)[\Gamma(H), Q(H)] + Fi[\Gamma(H), Q(H)] \\
&\quad - FPFi(1-q)[P(H), Q(H)].
\end{aligned}$$

(2.124)

But using Eq. (2.2), (2.25) and (2.77) one has:

$$PF = P\left(P + \frac{1}{q}Q\right) = P^2 = P,$$

(2.125)

and using Eq. (2.125), (2.25) and (2.77), then Eq. (2.124) becomes:

$$\begin{aligned}
C &= F^2((1-q+F^{-1})i[P(H), Q(H)] - F^{-1}Pi(1-q)[P(H), Q(H)]) \\
&= F^2(1-q+P+qQ-P(1-q))i[P(H), Q(H)] \\
&= F^2(1-q+q(P+Q))i[P(H), Q(H)],
\end{aligned}$$

(2.126)

or using Eq. (2.2), then

$$C = F^2 i[P(H), Q(H)].$$

(2.127)

This is a useful form for the inhomogeneous term in the lifted Jacobi equation, Eq. (2.120).[8]

In terms of the solution for $K(t)$, the lifted Jacobi field $J(t)$ for varying penalty parameter can first be written as:

$$J(t) = J(0) + \int_0^t dt' \frac{dJ(t')}{dt'}.$$

(2.128)

But according to Eqs. (2.109) and (2.105), one has:

$$\frac{dJ(t)}{dt} = U^\dagger(t)K(t)U(t),$$

(2.129)

and substituting Eq. (2.129) in Eq. (2.128), one obtains:

$$J(t) = J(0) + \int_0^t dt' U^\dagger(t')K(t')U(t').$$

(2.130)

Next consider the case in which the Hamiltonian is constant along a geodesic. The geodesic Eq. (2.76) then implies:

$$[G(H), H] = 0.$$

(2.131)

Also, using Eqs. (2.80), (2.26), (2.55), (2.2), (2.77) and (2.97), one has:

$$0 = \frac{dH}{dt} = G^{-1}\frac{dL}{dt} = iF[L, F(L)] = iF[L, P(L) + q^{-1}Q(L)]$$

$$= iF[L, P(L) + q^{-1}(1 - P)(L))]$$

$$= i(1 - q^{-1})F[L, P(L)]$$

$$= i(1 - q^{-1})F[P(H) + qQ(H), P(P(H) + qQ(H)]$$

$$= i(1 - q^{-1})F[P(H) + qQ(H), P(H)]$$

$$= i(q - 1)F[Q(H), P(H)]$$

$$= i(1 - q)F[P(H), Q(H)]$$

$$= i(1 - q)F[H, Q(H)]$$

$$= i(1 - q)F[H, G'(H)] = 0. \tag{2.132}$$

It then follows from Eqs. (2.120), (2.131) and (2.132) that if the Hamiltonian is constant, then again one obtains the conventional Jacobi equation, Eq. (2.121). Thus if the Hamiltonian is constant, and $J(0) = 0$ and $dJ(0)/dt = 0$, then in accord with Eq. (2.157) below, $J(t)$ is proportional to $dJ(0)/dt$ and therefore $J(t) = 0$. In this case it then follows that the geodesics for the lifted Jacobi equation for varying penalty parameter remain the same as for the conventional Jacobi equation and are the same for all values of the penalty parameter q.

The so-called geodesic derivative can be used to numerically determine geodesics which evolve from the identity $U(0) = I$ to a chosen unitary transformation U.[8] One first chooses a Hamiltonian $H(0)$ which produces $U = \exp(-iH(0)T)$ at some fixed time T along the geodesic for penalty parameter $q = 1$, in which case the Hamiltonian is constant. The parameter q can next be varied to produce a corresponding change in the initial Hamiltonian, and requiring that the lifted Jacobi field J vanish at $t = 0$ and $t = T$, this leads to the so-called geodesic derivative $dH_q(0)/dq$ derived below. Numerical integration then may produce a geodesic connecting the identity $U(0) = I$ and the chosen unitary transformation $U(T)$ for any penalty parameter q, provided that there are no intervening conjugate points.

To proceed then, the general lifted Jacobi equation, Eq. (2.120), for varied penalty parameter can be solved. First substituting Eq. (2.122) in Eq. (2.120), one has:

$$\frac{d}{dt}K = -iF([K, L] + [H, G(K)]) - C. \tag{2.133}$$

The corresponding homogeneous equation, for which $C = 0$ in Eq. (2.133), is then

$$\frac{d}{dt}K_s = -iF([K_s, L] + [H, G(K_s)]), \tag{2.134}$$

and it can be solved if it is first recast in vectorized form.[5,42] For any matrix

$$M = \begin{bmatrix} a_{11} & a_{12} & \cdots & a_{1n} \\ a_{21} & a_{21} & \cdots & a_{2n} \\ \vdots & \vdots & \cdots & \vdots \\ a_{m1} & a_{m2} & \cdots & a_{mn} \end{bmatrix}, \tag{2.135}$$

one defines the vectorized form of the matrix M by the column vector,

$$\text{vec}\, M = [a_{11} \cdots a_{m1}, a_{12} \cdots a_{m2} \cdots a_{1n} \cdots a_{mn}]^T, \tag{2.136}$$

with each column of the matrix M appearing beneath the previous one, arranged in a column vector. If one has a matrix equation:

$$C = AX + XB \tag{2.137}$$

for matrices A, B, C and X, then it can be shown that,[42]

$$\text{vec}\, C = [(I \otimes A) + (B^T \otimes I)]\text{vec}\, X. \tag{2.138}$$

It then follows that the homogeneous Eq. (2.134) can be written in vectorized form as follows:

$$\begin{aligned} \frac{d}{dt}(\text{vec}\, K_s) &= -iF\text{vec}[K_s L - L K_s + H G(K_s) - G(K_s)H] \\ &= iF\text{vec}[(L K_s + K_s(-L)) - (H G(K_s) + G(K_s)(-H))] \\ &= iF[(I \otimes L - L^T \otimes I)\text{vec}\, K_s - (I \otimes H - H^T \otimes I)\text{vec}\, G(K_s)] \\ &= iF[[(I \otimes L - L^T \otimes I) + (H^T \otimes I - I \otimes H)G]\text{vec}(K_s)] \\ &= iA\text{vec}\, K_s, \end{aligned} \tag{2.139}$$

where

$$A = F[(I \otimes L - L^T \otimes I) + (H^T \otimes I - I \otimes H)G], \tag{2.140}$$

and F and G are the vectorized forms of the superoperators F and G, respectively.[8] For example, the superoperator F acting on the matrix X can clearly be written as:

$$F(X) = \sum_j A_j X B_j, \tag{2.141}$$

for some matrices A_j and B_j. But one has[42]:

$$\text{vec}\, A_j X B_j = (B_j^T \otimes A_j)\text{vec}\, X, \tag{2.142}$$

and using Eq. (2.142) in Eq. (2.141), then,

$$\text{vec}\, F(X) = \sum_j (B_j^T \otimes A_j)\text{vec}\, X, \tag{2.143}$$

and therefore the vectorized form F of the superoperator F is given by:

$$F = \sum_j (B_j^T \otimes A_j)\text{vec} \,. \tag{2.144}$$

One can analogously obtain $\mathbf{G} = \text{vec}\, G$.

Evidently the solution to Eq. (2.139) is given in terms of the time-ordered exponential[43,44]:

$$\text{vec}\, K_s(t) = T\left(\exp\left(i\int_0^t A(t')dt'\right)\right)\text{vec}\, K_s(0)$$

$$= \left(1 + \sum_{n=1}^{\infty} \frac{i^n}{n!}\int_0^t dt_1 \cdots \int_0^t dt_n T(A(t_1)\cdots A(t_n))\right)\text{vec}\, K_s(0)\,, \tag{2.145}$$

where T denotes the time ordering operator. Equivalently, Eq. (2.145) is:

$$\text{vec}\, K_s(t) = \kappa_t \text{vec}\, K_s(0)\,, \tag{2.146}$$

where

$$\kappa_t = T\left(\exp\left(i\int_0^t A(t')dt'\right)\right)\,. \tag{2.147}$$

Next, it follows from Eq. (2.147) that:

$$\frac{d}{dt}\kappa_t = iA(t)\kappa_t\,. \tag{2.148}$$

Here κ_t is the propagator for the homogeneous form of Eq. (2.133), namely Eq. (2.134). The solution to Eq. (2.133) is then given by:

$$K(t) = \text{unvec}\left(\kappa_t \text{vec}\, K(0) - \kappa_t \int_0^t dr \kappa_r^{-1}\text{vec}\, C(r)\right)\,, \tag{2.149}$$

in which unvec unvectorizes,[8] namely, for a matrix M,

$$\text{unvec}(\text{vec}\, M) = M\,. \tag{2.150}$$

Equation (2.149) corrects Eq. (83) of Ref. 8.

Next, in order to obtain the propagator of the standard (unlifted) Jacobi field J_s, using Eq. (2.142), then Eq. (2.129) in vectorized form is:

$$\text{vec}\left(\frac{d}{dt}J_s\right) = (U^T \otimes U^{\dagger})\text{vec}\, K_s\,. \tag{2.151}$$

Substituting Eq. (2.146) in Eq. (2.151), then,

$$\text{vec}\left(\frac{d}{dt}J_s\right) = (U^T \otimes U^{\dagger})\kappa_t \text{vec}\, K_s(0)\,. \tag{2.152}$$

Next substituting Eq. (2.109) in Eq. (2.152),

$$\text{vec}\left(\frac{d}{dt}J_s\right) = (U^T \otimes U^{\dagger})\kappa_t \text{vec}\left(U(0)\frac{d}{dt}J_s(0)U(0)^{\dagger}\right)\,, \tag{2.153}$$

then since $U(0) = I$, Eq. (2.153) becomes:

$$\text{vec}\left(\frac{d}{dt}J_s\right) = (U^T \otimes U^\dagger)\kappa_t \text{vec}\frac{d}{dt}J_s(0). \tag{2.154}$$

Unvectorizing Eq. (2.154), then,

$$\frac{d}{dt}J_s = \text{unvec}\left[(U^T \otimes U^\dagger)\kappa_t\left(\text{vec}\frac{d}{dt}J_s(0)\right)\right]. \tag{2.155}$$

But since $J_s(0) = 0$, one has,

$$J_s(t) = \int_0^t dt' \frac{d}{dt'}J_s(t') \tag{2.156}$$

and substituting Eq. (2.155) in Eq. (2.156), then,

$$J_s(t) = \int_0^t dt'\text{unvec}\left[(U^T \otimes U^\dagger)\kappa_{t'}\left(\text{vec}\frac{d}{dt'}J_s(0)\right)\right]. \tag{2.157}$$

Next defining the propagator j_T that generates the standard unlifted Jacobi field at time T by:

$$J_s(T) = j_T\left(\frac{d}{dt'}J_s(0)\right), \tag{2.158}$$

then according to Eq. (2.157), one has:

$$j_T = \int_0^T dt'\text{unvec}(U^T \otimes U^\dagger)\kappa_{t'}\text{vec}. \tag{2.159}$$

It follows from Eqs. (2.149), (2.130), (2.156) and the homogeneous term having the same form as Eq. (2.158) that at time T the solution to the lifted Jacobi equation for varying penalty parameter is given by:

$$J(T) = j_T\left(\frac{d}{dt}J(0)\right) - \int_0^T dtU(t)^\dagger\left(\text{unvec}\,\kappa_t\left(\int_0^t dr\kappa_r^{-1}\text{vec}\,C(r)\right)\right)U(t). \tag{2.160}$$

Equation (2.160) corrects Eq. (85) of Ref. 8. Since $J(T) = 0$, Eq. (2.160) becomes:

$$\frac{d}{dt}J(0) = j_T^{-1}\left[\int_0^T dtU(t)^\dagger\left(\text{unvec}\,\kappa_t\left(\int_0^t dr\kappa_r^{-1}\text{vec}\,C(r)\right)\right)U(t)\right]. \tag{2.161}$$

Next, according to Eq. (2.90), the Hamiltonian H_q for penalty parameter q is such that:

$$\frac{d}{dq}H_q(0) = \frac{d}{dt}J(0), \tag{2.162}$$

and substituting Eq. (2.161) in Eq. (2.162), one obtains the so-called geodesic derivative[8]

$$\frac{d}{dq}H_q(0) = j_T^{-1}\left[\int_0^T dtU(t)^\dagger\text{unvec}\left(\kappa_t\left(\int_0^t dr\kappa_r^{-1}\text{vec}\,C(r)\right)\right)U(t)\right]. \tag{2.163}$$

Equation (2.163) corrects Eq. (86) of Ref. 8.

Equation (2.163) is next to be evaluated for penalty parameter $q = 1$. For $q = 1$, one has according to Eqs. (2.2), (2.25) and (2.55),

$$F = P + \frac{1}{q}Q = P + Q = G = 1, \quad q = 1, \tag{2.164}$$

and then according to Eq. (2.76) the Hamiltonian is constant,

$$\frac{dH}{dt} = 0, \quad q = 1. \tag{2.165}$$

Then Eq. (2.127) becomes:

$$C = i[P(H), Q(H)], \quad q = 1, \tag{2.166}$$

and then because of Eq. (2.165),

$$\frac{dC}{dt} = 0, \quad q = 1. \tag{2.167}$$

Therefore in Eq. (2.163) for $q = 1$ one has using Eq. (2.166),

$$Z \equiv \kappa_t \left(\int_0^t dr \kappa_r^{-1} \text{vec}\, C(r) \right) = \kappa_t \left(\int_0^t dr \kappa_r^{-1} i \text{vec}[P(H), Q(H)] \right), \tag{2.168}$$

and it follows that:

$$\frac{dZ}{dt} \equiv \left[\frac{d\kappa_t}{dt} \left(\int_0^t dr \kappa_r^{-1} \right) + 1 \right] i \text{vec}[P(H), Q(H)]. \tag{2.169}$$

But for $q = 1$ in Eq. (2.140), one has, according to Eqs. (2.164), (2.25) and (2.77), $F = 1$, $G = 1$ and $L = H$, and therefore:

$$A = [I \otimes H - H^{\mathrm{T}} \otimes I + H^{\mathrm{T}} \otimes I - I \otimes H] = 0, \quad q = 1. \tag{2.170}$$

Then substituting Eq. (2.170) in Eq. (2.148), one obtains:

$$\frac{d\kappa_t}{dt} = 0. \tag{2.171}$$

Also, according to Eq. (2.147) for $t = 0$, one has:

$$\kappa_t = 1. \tag{2.172}$$

Then substituting Eq. (2.171) in Eq. (2.169), one obtains:

$$\frac{dZ}{dt} \equiv i \text{vec}[P(H), Q(H)]. \tag{2.173}$$

Also, according to Eqs. (2.168) and (2.172),

$$Z(0) = 0. \tag{2.174}$$

Then combining Eqs. (2.168) and (2.174), one obtains for $q = 1$,

$$Z = it \text{vec}[P(H), Q(H)], \quad q = 1, \tag{2.175}$$

or using Eqs. (2.168) and (2.171), then,

$$\kappa_t \left(\int_0^t dr \kappa_r^{-1} \text{vec}\, C(r) \right) = it\text{vec}[P(H), Q(H)]\,, \quad q = 1\,. \tag{2.176}$$

Substituting Eq. (2.176) in Eq. (2.163), then for $q = 1$, one has:

$$\frac{d}{dq} H_q(0) = j_T^{-1} \left[\int_0^T dt U(t)^\dagger it[P(H), Q(H)] U(t) \right], \quad q = 1\,. \tag{2.177}$$

For $q \neq 1$, one has, substituting Eq. (2.127) in Eq. (2.163),

$$\frac{d}{dq} H_q(0) = j_T^{-1} \left[\int_0^T dt U(t)^\dagger \text{unvec}\, \left(\kappa_t \left(\int_0^t dr \kappa_r^{-1} \text{vec} F^2 i[P(H), Q(H)] \right) \right) U(t) \right]. \tag{2.178}$$

Equation (2.178) corrects Eqs. (86) and (87) of Ref. 8. Equations (2.177) and (2.178) give the so-called geodesic derivative[8] which is useful in the numerical determination of optimal geodesics.

3. Summary

In this review of some aspects of the Riemannian geometry of quantum computation, the Riemann metric, connection, curvature, geodesic equation, lifted Jacobi equation and geodesic derivative on the manifold of the $SU(2^n)$ group of n-qubit unitary operators with unit determinant were exposited in terms of the Lie algebra $su(2^n)$. The right-invariant metric is given by Eq. (2.24). The connection is given by Eq. (2.73). The sectional curvature is given by Eq. (2.83). The geodesic equation is given by Eq. (2.82). The lifted Jacobi equation on the $SU(2^n)$ manifold, for varying penalty parameter, is given by Eq. (2.133) and the solution is given by Eqs. (2.149) and (2.127). Also, the geodesic derivative is given by Eqs. (2.177) and (2.178). These equations are germane to investigations of conjugate points and the global characteristics of geodesic paths,[13,14] complexity of particular quantum algorithms and minimal complexity quantum circuits.[1,8,31] Other aspects of the Riemannian geometry of quantum computation can be found in Ref. 31.

Acknowledgment

The author wishes to thank K. K. Phau for the invitation to prepare this paper and present a lecture at the 5th Asia Pacific Workshop on Quantum Information Science held at Nanyang Technological University in Singapore, 25–28 May 2011 in conjunction with the Festschrift in honour of Vladimir Korepin.

This research was supported by the U.S. Army Research Laboratory.

References

1. M. A. Nielsen and I. L. Chuang, *Quantum Computation and Quantum Information* (Cambridge University Press, UK, 2000).
2. R. Montgomery, A tour of sub-Riemannian geometries their geodesics and applications, in *Mathematical Surveys and Monographs*, Vol. 91 (American Mathematical Society, Providence, 2002).
3. N. Khaneja, S. J. Glaser and R. Brockett, *Phys. Rev. A* **65**, 032301 (2002).
4. C. G. Moseley, Geometric control of quantum spin systems, in *Quantum Information and Computation II*, eds. E. Donkor, A. R. Pirich and H. E. Brandt, Proc. SPIE Vol. 5436 (SPIE, Bellingham, WA, 2004), pp. 319–323.
5. M. A. Nielsen, *Quantum Inf. Comput.* **6**, 213 (2006).
6. M. A. Nielsen *et al.*, *Phys. Rev. A* **73**, 062323 (2006).
7. M. A. Nielsen *et al.*, *Science* **311**, 1133 (2006).
8. M. R. Dowling and M. A. Nielsen, *Quantum Inf. Comput.* **8**, 0861 (2008).
9. J. N. Clelland and C. G. Moseley, *Differ. Geom. Appl.* **24**, 628 (2006).
10. B.-Y. Hou and B.-Y. Hou, *Differential Geometry for Physicists* (World Scientific, Singapore, 1997).
11. A. A. Sagle and R. E. Walde, *Introduction to Lie Groups and Lie Algebras* (Academic Press, New York, 1973).
12. L. Conlon, *Differentiable Manifolds*, 2nd edn. (Birkhäuser, Boston, 2001).
13. J. M. Lee, *Riemannian Manifolds: An Introduction to Curvature* (Springer, New York, 1997).
14. M. Berger, *A Panoramic View of Riemannian Geometry* (Springer-Verlag, Berlin, 2003).
15. J. Milnor, *Morse Theory* (Princeton University Press, Princeton, NJ, 1973).
16. B. C. Hall, *Lie Groups, Lie Algebras and Representations* (Springer, New York, 2004).
17. J. Farout, *Analysis on Lie Groups* (Cambridge University Press, Cambridge, 2008).
18. M. M. Postnikov, *Geometry VI: Riemannian Geometry, Encyclopedia of Mathematical Sciences*, Vol. 91 (Springer-Verlag, Berlin, 2001).
19. P. Petersen, *Riemannian Geometry*, 2nd edn. (Springer, New York, 2006).
20. J. Jost, *Riemannian Geometry and Geometric Analysis*, 5th edn. (Springer-Verlag, Berlin, 2008).
21. R. Wasserman, *Tensors and Manifolds*, 2nd edn. (Oxford University Press, Oxford, 2004).
22. M. A. Naimark and A. I. Stern, *Theory of Group Representations* (Springer-Verlag, New York, 1982).
23. M. R. Sepanski, *Compact Lie Groups* (Springer, Berlin, 2007).
24. W. Pfeifer, *The Lie Algebras su(N)* (Birkhäuser, Basel, 2003).
25. J. Stillwell, *Naive Lie Theory* (Springer, New York, 2008).
26. J. F. Cornwell, *Group Theory in Physics: An Introduction* (Academic Press, San Diego, 1997).
27. J. F. Cornwell, *Group Theory in Physics*, Vols. 1 & 2 (Academic Press, London, 1984).
28. W. Steeb and Y. Hardy, *Problems and Solutions in Quantum Computing and Quantum Information*, 2nd edn. (World Scientific, New Jersey, 2006).
29. S. Weigert, *J. Phys. A: Math. Gen.* **30**, 8739 (1997).
30. C. Reutenauer, *Free Lie Algebras* (Clarendon Press, Oxford, 1993).
31. H. E. Brandt, *Proc. Symposia Appl. Math.* **68**, 61 (2010).
32. H. E. Brandt, *Nonlin. Anal.* **71**, e474 (2009).

33. C. W. Misner, K. S. Thorne and J. A. Wheeler, *Gravitation* (W. H. Freeman and Company, New York, 1973).
34. P. D. Lax, *Commun. Pure Appl. Math.* **21**, 467 (1968).
35. R. Abraham and J. E. Marsden, *Foundations of Mechanics*, 2nd edn. (AMS Chelsea Publishing, American Mathematical Society, Providence, 2008).
36. D. Zwillinger, *Handbook of Differential Equations*, 3rd edn. (Academic Press, San Diego, 1998).
37. R. S. Kaushal and D. Parashar, *Advanced Methods of Mathematical Physics* (CRC Press, Boca Raton, 2000).
38. T. Miwa, M. Jimbo and E. Date, *Solitons* (Cambridge University Press, UK, 2000).
39. L. Debnath, *Nonlinear Partial Differential Equations* (Birkhäuser, Boston, 1997).
40. E. Zeidler, *Nonlinear Functional Analysis and its Applications IV: Applications to Mathematical Physics* (Springer-Verlag, New York, 1997).
41. H. E. Brandt, *Physica E* **42**, 449 (2010).
42. R. Horn and C. Johnson, *Topics in Matrix Analysis* (Cambridge University Press, UK, 1991), p. 255.
43. W. Greiner and J. Reinhardt, *Field Quantization* (Springer-Verlag, Berlin, 1996), p. 219.
44. S. Weinberg, *The Quantum Theory of Fields I* (Cambridge University Press, UK, 1995), p. 143.

Chapter 5

QUANTUM MECHANICS AND THE ROLE OF TIME: ARE QUANTUM SYSTEMS MARKOVIAN?

THOMAS DURT

Ecole Centrale de Marseille, Institut Fresnel,
Domaine Universitaire de Saint-Jérôme,
Avenue Escadrille Normandie-Niémen
13397 Marseille Cedex 20, France
thomas.durt@centrale-marseille.fr

The predictions of the Quantum Theory have been verified so far with astonishingly high accuracy. Despite of its impressive successes, the theory still presents mysterious features such as the border line between the classical and quantum world, or the deep nature of quantum nonlocality. These open questions motivated in the past several proposals of alternative and/or generalized approaches. We shall discuss in the present paper alternative theories that can be infered from a reconsideration of the status of time in quantum mechanics. Roughly speaking, quantum mechanics is usually formulated as a memory free (Markovian) theory at a fundamental level, but alternative, nonMarkovian, formulations are possible, and some of them can be tested in the laboratory. In our paper we shall give a survey of these alternative proposals, describe related experiments that were realized in the past and also formulate new experimental proposals.

Keywords: Quantum; Markovian; memory.

1. Introduction

In order to illustrate what we mean when we claim that, in its standard acception, quantum mechanics is assumed to be Markovian, let us consider the celebrated double slit experiment during which particles arrive one by one and create, after a sufficiently long collection time, a typical interference pattern.[1] It is commonly accepted that, provided their arrival times are sufficiently separated, all what these particles have in common is that they are described by the same wavefunction (that reflects their common preparation procedure). Otherwise, it is assumed that such particles are not correlated. Of course, quantum mechanics does not prohibit correlations between different particles, but in the typical situation considered here, it is usually assumed that these correlations are negligible and that the same wavefunction contains all the relevant information concerning the statistical behavior of the particles and the emergence of an interference pattern. When this implicit hypothesis of independence between successive preparations of a quantum system is satisfied we shall say that the quantum system is Markovian. The hypothesis that quantum systems are Markovian systems is, of course an *a priori* hypothesis, although it is a very natural hypothesis. Actually, it is well-known in cognitive sciences that it often occurs that the human mind, when it is confronted to the appearance of random phenomena, tries to find patterns and laws that would guide their appearance. This tendency can be at the source of serious cognitive illusions that have been studied in depth by psychologists (e.g., apophenia, illusory correlation[a] and so on; even magical thinking can be considered to belong to this category).

Now, it is not because a property accepted by many as being *a priori* true is very natural and has been confirmed in several circumstances that it should not be questioned.[b] This justifies our interest to investigate to which extent fundamental laws of nature (here we think to quantum mechanics) are Markovian theories, according to the definition given above.[c]

The problem that concerns us is in a sense very similar to the problem of deciding whether or not a series of bits 0 and 1 is random. The best that can be done to answer this question is to apply a series of tests aimed at revealing whether "serious" departures from the random behavior can be observed. Now, it is well-known that there exist well-conceived pseudo-random series that will succesfully

---

[a]A famous illustration of this tendency is[45] ... *the Gambler's fallacy, also known as the Monte Carlo fallacy (because its most famous example happened in a Monte Carlo Casino in 1913) or the fallacy of the maturity of chances, which is the belief that if deviations from expected behavior are observed in repeated independent trials of some random process, future deviations in the opposite direction are then more likely* ... Opposite to this illusion, one finds an effect that has been baptized in cognitive sciences under the name of "clustering illusion".[2] ... *The clustering illusion refers to the tendency erroneously to perceive small samples from random distributions to have significant "streaks" or "clusters", caused by a human tendency to underpredict the amount of variability likely to appear in a small sample of random or semi-random data due to chance. The clustering illusion was central to a widely reported study by Gilovich, Robert Vallone and Amos Tversky. They found that the idea that basketball players shoot successfully in "streaks", sometimes called by sportcasters as having a "hot hand" and widely believed by Gilovich et al.'s subjects, was false. In the data they collected, if anything the success of a previous throw very slightly predicted a subsequent miss rather than another success* ...

[b]As L. de Broglie wrote[3]:

"L'histoire des Sciences montre que les progrès de la Science ont constamment été entravés par l'influence tyrannique de certaines conceptions que l'on avait fini par considérer comme des dogmes. Pour cette raison, il convient de soumettre périodiquement à un examen très approfondi les principes que l'on a fini par admettre sans plus les discuter." ...

In other words, commonly accepted truths are questionable too.

[c]One should be careful not to confuse our acception of the label "Markovian" (respectively "non-Markovian") with the acception that it carries in other approaches. For instance, in the framework of stochastic quantum dynamics, so-called nonMarkovian effects are likely to occur when a quantum system is coupled to an environment in such a way that the environment will "remember" the instantaneous state of the system for a time that is not short in comparison to the typical time of evolution of this system.[4,5] Also, we do not have in mind here what is called "Markovian process", which is defined in the framework of classical probability theory.[6-8]

pass all the standard tests despite of the fact that there exists a deterministic algorithm that allows to generate them on the basis of a short key (the "seed"). Sometimes, there exist specific tests aimed at revealing the nature of the seed and of the algorithm, so that one can decide whether these pseudo random series are, indeed, not random, but most often these tests are not known. A famous example is provided by the "RANDU" generator that was massively used in the 60's to generate random numbers, until one could conceive a test that explicitly revealed its pseudo-random nature.[d] Considered so, the question "Is a certain series random?" is undecidable.

Similarly, the question "Are quantum systems Markovian?" is undecidable. The best that we can do is to conceive tests that make it possible to reveal specific departures from the Markovian behavior, but it is only our own imagination that limits the number of tests that can be conceived[e] ...

The main goal of our work is to describe various models which lead to the prediction of nonMarkovian effects for quantum systems, and the related experiments and experimental proposals that can be considered as attempts, in very specific situations, to quantify the departure from the Markovian paradigm.

The paper is organized as follows. In the second section, we sketch several experiments that were aimed at testing the Markovian nature of quantum probabilities, in the nonrelativistic regime. We devote particular attention to the Paris–Nord experiment[9] for which we describe certain statistical tests that were realized by us recently in order to scrutinize in depth the possibility of a nonMarkovian signature in the quantum signal collected in Paris–Nord's atomic interferometer.[10] In the third section, we consider a recent proposal by Lee Smolin called "Real Ensemble Interpretation of Quantum Mechanics"[11] that we compare with Rupert Sheldrake's "Morphic Field theory".[12] We formulate an experimental proposal aimed at testing hypotheses inspired by Smolin and Sheldrake's models. The last section is devoted to the conclusions.

2. Do Quantum Dice Remember? — Experimental Tests

We shall now shortly describe four approaches in which departures form the Markovian paradigm are supposed to be possible, and also sketch experimental tests that were motivated by them.

These are:

(3.1) Bohm–Bub's model,,[13] and the Papaliolos experiment[14]
(3.2) Buonomano's model[15] and the Summhammer experiment[16,17]

---

[d]It is also notoriously known among cryptographists that the Matlab random number generator can be broken one knows enough consecutive bits of one of its random series.

[e]Actually, in a next section, we show by explicit examples that certain tests used for deciding whether a series of bits is random can also be used to gain information about the Markovian nature of quantum systems.

(3.3) the so-called hidden measurement approach[18,19] and an experiment realized in 1999 and 2000 in Paris–Nord in order to test it.[10]

(3.4) Shnoll's model[20] and recent experimental results that were collected with a Quantum Random Number Generator (QRNG) in Brussels.[21–23]

2.1. *Bohm–Bub's model and the Papaliolos experiment*

The experiment realized by Papaliolos in 1967[14] was aimed at testing the relevance of Bohm–Bub's theory.[13] In this theory, which describes the interaction between a quantum system and a measuring device, it was assumed that hidden variables exist, which complete the description of the state of a quantum system provided by the wavefunction. The authors suggested the hypothesis according to which these hidden variables would randomize in a time comparable to h/kT, where h, k and T are the Planck constant, the Boltzmann constants and the temperature of the environment respectively.[24,25] If we apply the Bohm–Bub theory to the case of a two level system, in order to describe the passage of a photon through a polarizing device, and if we assume that the hidden variable which hypothetically determines the result of a measurement does not randomize instantaneously, but remains "frozen" during a typical memory time, and is "carried" by the photon, nonstandard correlations must appear (departure from the Malus law[14,26]) between successive measurements carried out within a very short time interval. In order to check these effects, Papaliolos let pass low intensity light pulses prepared in a given polarization state through two successive, very close to each other, polarizers. If the distance between these two polarizers is smaller than the randomization time of the hidden variable times the speed of light, new correlations are predicted by the Bohm–Bub theory. Papaliolos did not observe any such effect even for extremely short distances,[14] and considered this negative result as the proof of the nonexistence of Bohm–Bub-like variables.

2.2. *Buonomano's model and the Summhammer experiment*

The goal of Summhammer's experiment, realized in 1985,[15–17] was to check the validity of the local hidden variable theory of Buonomano.[15,17] The idea of Buonomano was the following: the violation of Bell's inequalities[27,28] shows that particles are in some manner informed about the setting of distant apparatuses. It is possible to conciliate this phenomenon with Einsteinian causality provided we assume that, (a) particles are localized, and (b) this process of information is realized by the particles themselves. The particles would spread in their surroundings (thanks to a kind of quantum "flavor" of "pheromon") some information about the properties of the external world that they encounter along their way. By doing so, they could (*a posteriori*) inform the next incoming particle about the obstacles along the way. This next particle could inform the following particle about the nature of the obstacles to be met in the future and so on, untill the information reaches the source. The aim of this hypothesis was to explain how sources of entangled particles could "foresee" what would be the configurations of distant measuring devices. In order to check these ideas, Summhammer used a low-intensity neutronic interferometer in which nonlocal properties appear: a single neutron is spread among two arms which are separated by a distance much larger than the coherence length of the neutron. Afterwards, the bundles originating from both arms are superposed, and interfere according to the standard quantum predictions. Thanks to a shutter, Summhammer opened and closed at random one arm within the interferometer in order to permit a sudden and random switching from interfering to noninterfering conditions and vice versa. Then, if the theory of Buonomano was right and that a particle could "collect" information stored by the previous particles about the internal configuration of the interferometer, it would get misled because in the meanwhile the experimental set up would have been changed. If Buonomano's hypothesis would have been correct, the genuinely predicted probabilities would have been disturbed by memory-effects. Summhammer realized the experiment[16] and did not observe any discrepancy with standard quantum predictions, even directly after the opening or the closing of the shutter, which constitutes a clear negative result and proves unambiguously the nonexistence of the Buonomano type hidden-variables.

2.3. *Hidden measurement approach and the Paris–Nord experiment*

D. Aerts proposed in 1982[18] a hidden variable model in which the hidden variables are present at the level of the detector (and not at the level of the quantum system as is postulated in the majority of hidden variable theories), let us call them hidden measurement variables. As is shown in references,[19,29,30] one could test the hypothetical existence of such variables provided they exhibit memory effects. Such effects were not tested in previously mentioned experiments (with light and neutrons) but they were tested (in 1999) with neutral atomic beams in the Laboratoire de Physique des Lasers de l' Université de Paris–Nord.[10] Essentially, the goal of the experiment was to test the possibility of a memory effect inside the detector. To conceive an experiment which allows us to check the presence of hidden variables inside the detector (hidden measurement variables), it is sufficient to reconsider Papaliolos experiment, and to permute in it the roles played by the system under measurement and the measuring apparatus. We must thus consider the probability of, for instance, measuring successively a dichotomic property on two quantum systems initially prepared in a same state. The assumption of the existence of hidden measurement variables characterized by a measurable memory time leads to the prediction of results which will differ from standard predictions.[10] Roughly speaking, we expect, if hidden measurement variables exist and vary continuously in time, that successive measurements performed rapidly enough on systems prepared in the same superposition state will exhibit a tendency to provide successive identical results which is higher than what we could expect in the case of independent, non-correlated measurements. This is analog to the so-called bunching experiments of quantum optics[31–34] in which, because of their bosonic nature, photons were shown to have a tendency to arrive together. In this case, the distribution of the arrival times was shown to be a surPoissonian distribution. In our case, photons are replaced by atoms. The standard distribution of arrival times of the atoms being Poissonian, we also expect to observe deviations from the Poissonian distribution. An explicit model was developed in Ref. 10 that shows that the deviation from the standard, Poissonian, value of the normalized square root deviation of the distribution of delay times between successive detections collected in an interferometer is proportional to the memory time of the detector. We shall now describe the experiment that was realized by sending 2.6 millions of successive atoms in the interferometer and establishing the statistical distribution of the delay times (times sery) between them and rederive the main features of our model. As we shall now comment, no deviation from standard, memory-free, distributions could be revealed, up to now.

2.3.1. The Poissonian paradigm

Schematically, the device consists of a source that emits metastable excited atoms.[10] These atoms are collimated and arrive in the region of detection in which an electric field induces a stimulated decay of the excited atomic states that is accompanied by the emission of a photon. This will happen with a probability P_Q (P_Q depends on the excited state in which the atom was originally prepared and on the intensity of the electric field that stimulates the desexcitation). This photon is then detected through photo-electric effect in a channel electron amplifier (channeltron). In first approximation, this detection will occur with a probability P_D that is of the order of the product of the efficiency of the detector with its solid angular opening relatively to the region where the desexcitation occurs divided by 4π. If, at the time t_0, an atom passes at the level of the photon detector, and that successive passages are independent (this constitutes the fundamental hypothesis of Markovicity in our treatment), the probability that an other atom passes during a short time-interval t_0, $t_0 + dt$ is thus, up to the first order in dt, equal to dt/T, where T is the average time-delay between the passage of two successive atoms. We have thus that the probability of no-detection $\mathcal{P}_{\text{neg}}(t_0, t_0 + n \cdot dt)$ during a time $n \cdot dt$, where n is a large integer, is equal, in first approximation, to the product of the probabilities of no-detection for all intermediate dt intervals: $\mathcal{P}_{\text{neg}}(t_0, t_0 + n \cdot dt) = \Pi_{i=1}^{i=n}(1 - P_Q P_D \cdot dt/T)$. If we consider the limit of this expression for extremely short intervals dt, we find that $\mathcal{P}_{\text{neg}}(t_0, t_0 + n \cdot dt) = \exp(-n \cdot P_Q P_D \cdot dt/T)$. From this expression, it is straightforward to deduce that we have again a Poissonian distribution $\mathcal{P}(t)$ for the time-delays. The probability $\mathcal{P}(t) \cdot dt$ that the next photon is detected inside the interval $[t_0 + t, t_0 + t + dt]$ when a first photon is detected at time t_0 is equal to the product of the probability $\mathcal{P}_{\text{neg}}(t_0, t_0 + n \cdot dt)$ that no-detection occurs between t_0 and $t_0 + t$ with $P_Q P_D \cdot dt/T$, the probability of a detection inside the interval $[t_0 + t, t_0 + t + dt]$: $\mathcal{P}(t) = (1/T)P_Q P_D \cdot \exp(-P_Q P_D \cdot (t/T))$.

2.3.2. *Departures from the Poissonian paradigm*

It could happen that successive detections are no longer independent, for one or another reason, in which case departures from the Poissonian distribution could be observed. A very convenient parameter aimed at measuring the departure from the Poissonian statistics is provided by the normalized mean square root deviation g, so to say, the ratio between the mean square root deviation and the mean value of the distribution of the time-delays between the detections of two successive photons. When no memory effect is present, the distribution is Poissonian and this parameter g is equal to unity. When detections tend to arrive together, g gets larger than one (bunching), and when a detection is followed by a dead time, it gets smaller than one (antibunching). In another paper we described how the existence of short range memory effects inside the detector itself could also be revealed by the departure from unity of g. We showed that when the sensibility of the detector gets enhanced (inhibited) for a short time τ directly after a measurement the departure from unity of the normalized mean square root deviation is proportional to $+ (-) \tau/\tilde{T}$ where $\tilde{T}$ represents the average time $T/P_Q P_D$ between successive detections (with $\tau < \tilde{T}$). We estimated the normalized mean square root deviation from a sample of $N = 2.6 \cdot 10^6$ time delays and checked that this value was well equal to one, up to typical statistical fluctuations that can be predicted thanks to the law of large numbers in the case of a Poissonian distribution. This shows that the distribution of time-delays is well Poissonian as far as we can know and that if a memory time τ exists it must be such that the departure from one of g_N, the measured value of g, stays inside the statistical fluctuations. Let us denote by $t_{\bar{N}}$ (resp. $(t^2)_{\bar{N}}$) the variable that we obtain by an averaging process after having measured N times the value of t (resp. t^2): $t_{\bar{N}} = (t_1 + t_2 + \cdots + t_N)/N$ (resp. $(t^2)_{\bar{N}} = (t_1^2 + t_2^2 + \cdots + t_N^2/N)$). Let us denote by $g_{\bar{N}}$ the estimation of the mean squared root deviation g that we obtain from this sample of N data: $g_{\bar{N}}^2 = (t^2)_{\bar{N}}/t_{\bar{N}}^2 - 1$. By assuming that successive detections are independent and by making use of the properties of the Poissonian distribution, we get that the standard deviation of $g_{\bar{N}}$, denoted $\sigma(g_{\bar{N}})$ can be upperly bounded as follows:

$$\sigma(g_{\bar{N}}) = \sigma\left(\frac{(t^2)_{\bar{N}}}{t_{\bar{N}}^2}\right) \leq \frac{\sigma((t^2)_{\bar{N}})\langle t_{\bar{N}}\rangle^2 + 2\langle t_{\bar{N}}\rangle\sigma(t_{\bar{N}})(t^2)_{\bar{N}}}{\langle t_{\bar{N}}\rangle^4}$$

$$= \frac{4 + \sqrt{8}}{\sqrt{N}} \simeq \frac{6.83}{\sqrt{N}}.$$

Note that it is consistent to assume Poissonicity because we are presently working at the lowest order (order zero) in $\tau/\tilde{T}$. After having accumulated $N = 2.6 \cdot 10^6$ measured values of the time delay between successive detections, we got that $g_{\bar{N}} = 1.0001$. According to the law of large numbers, the extent of the interval of values of $g_{\bar{N}}$ centered around one which ought to be occupied in 95% of the cases is equal to $\sqrt{1,96/N} \cdot \sigma \approx \sqrt{1,96/N} \cdot 6.83$ so to say, to $5.93 \cdot 10^{-3}$ when $N = 2.6 \cdot 10^6$, in agreement with the observed value. Now, it could occur that $\tau/\tilde{T}$ is not equal to zero but is so small that the departure from one of g_N belongs to the margin of statistical fluctuations. This is possible provided $\tau < 5.93 \cdot 10^{-3} \cdot T$. The average time between two detections being equal to 0.5 ms, the conclusion of this analysis was thus that, even if some memory time (or eventually some dead-time) characterizes the detection process, such a time may not exceed $5.93 \cdot 10^{-3} \cdot 0.5$ ms so to say $\sim 3 \cdot 10^{-6}$ seconds. This preliminary analysis, realized in 2000, suggested strongly that no departure from the Poissonian distribution could be observed in the temporal distribution of "clicks" obtained in Paris–Nord's atomic interferometers.

2.3.3. *Recent results*

The negative result mentioned by us in the previous section, so to say that no departure from the Markovian behavior was observed is not a definitive result; it is only a part of the story. Effectively we could imagine different scenarios for which this parameter is not representative of the existence of memory effects, for several reasons. For instance, we assumed implicitly in the model mentioned in the previous section that the memory effect of the detector is always similar to itself throughout time: either its sensitivity gets enhanced after a detection (bunching), either it gets inhibited (anti-bunching) but it could be that sometimes it gets enhanced and sometimes it gets inhibited (because of some hidden drift effect) in such a way that in average the global distribution is still Poissonian.

The existence of effects of this kind could be revealed in the case where the drift period is stable by a Fourier analysis of the distribution of time delays. Recently, we investigated this hypothesis. Once more, no such effect was observed: the Fourier distribution shows no departure from the distribution of a Poissonian distribution inside the limit of validity provided by the law of large numbers.

Even in the case that the period of the drift is not stable, one could measure a nonstandard memory effect of the type described previously (footnote a), according to which "streaks" of higher sensitivity as well as streaks of lower sensitivity appear during the measurement process, thanks to the so-called persistency parameter that is aimed at revealing the tendency of fluctuations to persist during a certain time.

Historically, persistency[35] was firstly observed by the English statistician Hurst who studied the statistical data collected throughout milleniums by Egyptian historians about the fluctuations of the level of the river Nile. It was indeed considered at that time that the level of the river Nile (that conditioned the food production of the whole Empire) was representative of the opinion of the gods about the pharaonic regime. This explains why the values of the yearly fluctuations of the river Nile were measured and collected by the court archivists with great care and accuracy. These fluctuations exhibit an effect called the Joseph effect that consists of persistency in the time-series. The word Joseph refers to the bible where the appearance of a periodicity of six years in dry and fertile years is mentioned. Such an effect occurs when the sign of the departure from the average value of the time-delay persists over M successive measures ($M = 1, 2, \ldots$). The strength of this effect is measured by the persistency parameter which is equal, for a sample that contains N realizations of a random variable V_i $(i = 1, \ldots, N)$, to

$$\frac{1}{N - 2M + 1} \cdot \left\{ \sum_{j=M\cdots N-M} \left(\frac{1}{M \cdot \sigma(V)^2} \left(\sum_{i=j-M+1}^{j} (V_i - V_{\bar{N}}) \right) \right) \right.$$
$$\left. \cdot \left(\sum_{i=j+1}^{j+M} (V_i - V_{\bar{N}}) \right) \right\}.$$

When the persistency parameter is positive (negative), persistency (antipersistency) occurs. When the signal is Markovian, the persistency parameter does not significatively differ from zero.

Recently, we investigated the possibility that persistency is present inside the signal collected in Paris–Nord. In order to do so, we generated a random bit by converting the measured data (that is the Poisson distributed time T between two clicks at the output of the interferometer) into the bit value 0 in the case that this time is comprized between zero and the median of T and into the bit value one otherwise. Then, we estimated the persistency parameter and found no contradiction with the law of large numbers. All these results confirm the unobservability of hypothetical nonstandard effects in the quantum signal.

2.4. *Shnoll's hypotheses and the Brussels experiment*

According to Shnoll,[20] even in the case that the law of large numbers is verified, so that the statistical data collected during a very long period will obey the normal distribution, departures from the normal distribution exhibit nonMarkovian memory effects of a very peculiar nature. The main hypotheses of Shnoll regarding these memory effects are the following:

(a) Departures from the expected behavior (so to say from the normal distribution) are likely to repeat themselves (or to appear in "streaks" according to the terminology introduced in footnote a).

(b) The corresponding likelihood (or propensity) exhibits a peak for short times (this is a short-range memory effect).

(c) This likelihood also exhibits peaks for time periods equal to (i) the rotation period of Earth around itself (24 hours) (ii) the rotation period of Moon around Earth (more or less 28 days) (iii) the rotation period of Earth around the Sun (one year).

(d) Moreover, this memory effect is assumed by Shnoll to characterize all *NATURAL* (e.g., physical) sources of randomness. According to Shnoll, this comprises fluctuating biological, chemical, electronic, thermic and even quantum systems that can be found in nature; for instance, Shnoll assumed that the data collected regarding the statistics of decay times of unstable, radio-active, quantum systems exhibits memory effects that would obey hypotheses (a, b, c).

In a sense, one could say that Shnoll's hypothesis expresses the belief that if deviations from expected behavior are observed in repeated independent trials of some random process, future deviations in the same direction are then more likely, which is to some extent a variation on the theme provided by the aforementioned "Clustering Illusion" (footnote a).

In order to test Shnoll's hypotheses (a, b, d), we considered the random data extracted from an ultrafast QRNG that was conceived at the Université Libre de Bruxelles (ULB).

It is not our goal to describe here in detail the results of our analysis that will be published elsewhere,[21,22] nor to describe the internal functioning of the ULB QRNG.[23,36] All what we need to say here is that this optical QRNG, based on vacuum fluctuations, allowed us to produce normally distributed random numbers at a rate of one gigahertz. We conceived an algorithm that measures the distance between departures from the normal distribution exhibited by 1000 successive histograms each of which had been established on the basis of a series of 1000 detections. In a first time, our analysis revealed a significant deviation from the law of large numbers,[21,22] and this effect survived when we repeated our analysis considering several samples of 10^6 detections. Now, it appeared that when the QRNG was properly isolated from its electromagnetic environment, thanks to a Faraday shielding, the effect disappeared. We shall comment these results in the conclusion.

3. Sheldrake and Smolin's Models, and a Related Experimental Proposal

3.1. *Sheldrake's model*

Rupert Sheldrake[12] is the father of the so-called morphic field hypothesis according to which ... *there is a field within and around a morphic unit which organizes its characteristic structure and pattern of activity* .... In particular, due to the action of this hypothetical field, the evolution of living organisms at the surface of the earth would be accelerated by a kind of cosmic memory that would gather and disseminate (nonlocally) at the scale of the planet the teachings brought by experiences lived by individual organisms. That such a field exists at the level of living organisms is still an open question. Actually, Sheldrake's concept has little support in the mainstream scientific community and is rather considered to belong to pseudo-sciences and scientific heresy. It is commonly considered that Sheldrake's concept is currently unfalsifiable and therefore beyond of the scope of scientific experiment.

What interests us in particular in the present context is that R. Sheldrake also formulated the hypothesis according to which the de Broglie frequency (that is equal to the mass-energy of a quantum object divided by the Planck constant) expresses the level of self-memory of this object. Inertia, or resistance to a change of position, that is also proportional to the mass-energy would reveal the tendency of an object to "repeat itself", or to stay equal to itself. Considered so, one should expect that any quantum system is characterized by an intrinsic "memory time"[12] equal to h/mc^2, which is the Compton period of the object. Obviously, this idea directly concerns memory effects in the quantum regime which is the main topic of our paper. Actually, a model due to Smolin presents interesting analogies with Sheldrake's model as we shall discuss now.

3.2. *Smolin's model*

Lee Smolin formulated recently a new interpretation of the quantum theory called "Real Ensemble Interpretation of Quantum Mechanics".[11] It is out of the scope of the present paper to describe all subtleties of this model. Instead we shall directly treat a simple case, a two-level quantum system, for instance the polarization degree of freedom of a photon of given frequency. A first assumption of Smolin is that the quantum theory describes an ensemble of N identical systems, all of them initially prepared in the same state. In the present case, this ensemble would consist of all photons of the universe that are prepared in the same state of polarization and have the same frequency as the photon that interests us. N is of course a huge number. A second assumption is that nothing allows us to distinguish these photons (even their location is irrelevant according to Smolin, due to the fact that his model is intrinsically nonlocal). A third assumption is that there exists a privileged basis of the Hilbert space assigned to the system (in the present case, a basis that consists of two orthonormal polarization states $|a\rangle$ and $|b\rangle$) such that at any time the value of the polarization is either a or b. In other words, the polarization, in this privileged basis is what John Bell called a beable, and what is called an element of reality in EPR paper.[37] This is similar to the position of massive particles in the Bohmian picture that is assumed to be a "real", objective, property of the particle, at any time. A fourth ingredient of Smolin's model is that two extra-beables,[f] the phases $e^{-i\phi_a}$ and $e^{-i\phi_b}$ are respectively assigned to a and b. Without entering into details, the model is built in such a way that these phases are uniquely and unambiguously defined at any time.

A fifth ingredient is the idea that the beables of individual systems tend to copy the beables of other individual systems, and that the tendency to copy the beables of other systems is an increasing function of their relative weight. For instance, in the present case, where at any time there are N_a systems with the polarization "hidden" state a and N_b systems with the polarization "hidden" state b, the rate of jump from a to b (respectively b to a) is proportional to $\sqrt{N_b/N_a}$ (respectively $\sqrt{N_a/N_b}$). In the case that the quantum Hamiltonian that describes the polarization corresponds, when expressed in the privileged basis, to the two times two matrix

$$H^{\text{polarization}} = \hbar \begin{pmatrix} \omega_a & R_{ab}e^{i\delta_{ab}} \\ R_{ab}e^{-i\delta_{ab}} & \omega_b \end{pmatrix},$$

---

[f]This is a novelty regarding previous hidden beable models.[38,39]

where ω_a, ω_b, R_{ab} and δ_{ab} are real parameters R_{ab} is real positive), the precise dynamical laws postulated by Smolin are the following:

(i) The individual rate of jump from a to b is equal to $2\sqrt{N_b}/\sqrt{N_a}\sin^+(\phi_a - \phi_b + \delta_{ab})R_{ab}$ where the $\sin^+$ is defined to be equal to $\sin(\phi_b - \phi_a - \delta_{ab})$ when the last quantity is positive and to zero otherwise.

Correspondingly, the individual rate of jump from b to a is equal to $2\sqrt{N_a}/\sqrt{N_b}R_{ab}\sin^+(\phi_a - \phi_b + \delta_{ab})$ which is defined in a similar fashion.

The average or global rates must be weighted by ρ_a (ρ_b) respectively, where $\rho_a = N_a/N$ and $\rho_b = N_b/N$ are the relative frequencies of appearance of the polarization values a and b.

(ii) The temporal evolution of the phases ϕ_a and ϕ_b are:

$$\frac{d\phi_a}{dt} = \omega_a + \sqrt{\frac{N_b}{N_a}}R_{ab}\cos(\phi_a - \phi_b + \delta_{ab})$$

and

$$\frac{d\phi_b}{dt} = \omega_b + \sqrt{\frac{N_b}{N_a}}R_{ab}\cos(\phi_a - \phi_b + \delta_{ab}).$$

In accordance with the Born rule, the instantaneous quantum state of polarization is $|\Psi\rangle = \sqrt{\rho_a}e^{-i\phi_a}|a\rangle + \sqrt{\rho_b}e^{-i\phi_b}|b\rangle$.

Then one can check that for instance

$$\frac{d(\sqrt{\rho_a}e^{-i\phi_a})}{dt} = \frac{d\sqrt{\rho_a}}{dt}e^{-i\phi_a} + \sqrt{\rho_a}\frac{de^{-i\phi_a}}{dt}$$

$$= \frac{\frac{d\rho_a}{dt}}{2\sqrt{\rho_a}}e^{-i\phi_a} + \sqrt{\rho_a}\cdot(-i)$$

$$\cdot\left(\omega_a + \sqrt{\frac{N_b}{N_a}}R_{ab}\cos(\phi_a - \phi_b + \delta_{ab})\right)\cdot e^{-i\phi_a}$$

$$= \sqrt{\rho_a}\cdot(-i)\cdot\omega_a e^{-i\phi_a} + R_{ab}e^{-i\phi_a}$$

$$\cdot\left(+\frac{\rho_a\cdot\frac{\sqrt{\rho_b}}{\sqrt{\rho_a}}}{2\sqrt{\rho_a}}2\sin(\phi_a - \phi_b + \delta_{ab}) - i\sqrt{\rho_b}\cos(\phi_a - \phi_b + \delta_{ab})\right)$$

$$= \frac{1}{i}\cdot(\omega_a\cdot\sqrt{\rho_a}e^{-i\phi_a} + \sqrt{\rho_b}\cdot e^{-i\phi_a}\cdot R_{ab}e^{i(\phi_a - \phi_b + \delta_{ab})}), \tag{1}$$

in accordance with Schrödinger equation $i\hbar(\partial/\partial t)|\Psi\rangle = H^{\text{polarization}}|\Psi\rangle$. A similar property holds for what concerns the b component.

3.3. *Unfalsifiability of Sheldrake and Smolin's models — New experimental proposals*

In the framework of Bohm–Bub theory, Cerofolini[40] showed independently of Sheldrake's model that it is impossible to measure memory effects at the level of quantum systems when they are of the order of the Compton period. This is easy to understand, because for instance if light is used to probe a massive particle, we would have, in order to reach a temporal accuracy of the order of the Compton wavelength, to manipulate photons so energetic that they would be able to create new particles. Therefore Sheldrakes hypothesis about a quantum memory time of the order of the Compton period is unfalsifiable.

Besides, Smolin's model is also unfalsifiable, in the same sense that Bohm's theory is unfalsifiable, because it leads to exactly the same predictions as standard quantum mechanics.[g]

Actually, we treated Sheldrake's model and Smolin's model in the same section because they present very strong analogies: both models predict a tendency in nature to copy pre-existing, "objective", structures (shapes, beables). Sheldrake's model emphasizes the temporal aspect of this process, why in Smolin's model, it is the nonlocal nature of the process that is emphasized.

Now, it is clear that all the models discussed so far (hidden measurement approach, Shnoll's hypothesis, Sheldrake model, Smolin's model), if they have any degree of truth, must be considered as approximations to a more general theory which we do not know yet. What is interesting is that there are striking analogies between all these models. In all approaches, it is predicted that on certain scales, the independence of the outcomes of certain experiments is no longer guaranteed and it is predicted that certain outcomes are likely to be measured in "streaks", during a characteristic time that depends on the model.[h]

The nonstandard correlation scale is sometimes defined in the framework of the model (like in Sheldrake's model[12] where it is assumed that the typical memory time of the morphic field is of the order of Compton period, or like in the Bohm–Bub model[13] where this time is assumed to be of the order of Planck's constant divided by kT where k is Boltzmann's constant and T the temperature of the system of interest).

In Smolin's model, the typical spatial scale is the size of the cosmos, because the model is intrinsically nonlocal and it is assumed that the distance between equivalently prepared undistinguishable particles does not matter.

Now, nothing forbids to formulate more restrictive hypotheses regarding the typical scales that characterize the previous effects. For instance one could imagine that the morphic field is characterized by a mesoscopic memory time that is accessible experimentally; similarly, the tendency in Smolin's model of (the beable of) an individual system to copy (the beable of) another one could be a decreasing function

---

[g]Of course, we do not have in mind here theories *à la* Valentini[42–44] where it is assumed that departures from the Born rule are possible.

[h]In Smolin's model, one can predict for instance that if a statistical fluctuation occurs such that there are more systems that are found in the polarization state a than $(\sqrt{\rho_a})^2$ $((N_a/N) > \rho_a)$, the rate of transition from b to a will be higher than the quantum predicted rate during a certain period. Of course in order to check this prediction one ought to measure all undistinguishable systems prepared equivalently everywhere in the universe which is impossible, in agreement with the fact that Smolin's model is unfalsifiable.

of the spatio-temporal (Minkoskian) distance that separates the measurements of these beables.

This suggests a new experiment during which one checks whether the outcomes of measurements of polarizations of identically prepared systems are independent.

This experiment ought to be realized in different bases because *a priori* we do not know which would be the privileged basis. It would also be necessary to tailor the Hamiltonian in such a way that the transfer rate between the beables is not equal to zero during the experiment (that is $R_{ab} \neq 0$). It is possible to do so thanks to birefringent media possibly combined with optical wave plates. For instance, if the birefringent medium is such that two orthogonal linear polarizations propagate at different speeds and that the prefered basis is the basis of circular polarizations, then, the Hamiltonian is obviously nondiagonal in the beable basis.

One could investigate the existence of hypothetical temporal correlations by measuring successively a series of equivalently prepared photons, but one could also investigate whether there exist (nonlocal) correlations between the outcomes obtained from two equivalent devices placed in the same lab, in two vicine locations.

4. Conclusions

Experimental violations of Bell like inequalities[27,28] are not the single tests that can be performed in order to discriminate between hidden variable theories and standard quantum mechanics. As we have discussed in the present paper, well-chosen tests can for instance be realized in order to reveal whether hidden elements of reality would exist, provided they are characterized by a measurable, mesoscopic, memory time. The majority of the experiments described in this work confirmed the standard features of commonly accepted theories such as special relativity and quantum mechanics: the physical world "lives here and now", it does not remember, and supraluminal transmission of information is not possible. Now, strictly speaking, the existence of such effects is not absolutely impossible. One could always argue that a memory time exists but that it is too short to be observed experimentally. Therefore, rather than showing definitively the nonexistence of hidden times in quantum mechanics, the experiments mentioned in the present paper provide upper bounds on the typical memory times that could possibly exist. In summary, these constraints are the following:

- the Papaliolos experiment[14] (with polarized photons) showed that if the polarization of individual photons exhibits a randomization time, this time is certainly less than $2.4 \ 10^{-14}$ s.
- the Summhammer experiment[16] (with a neutronic interferometer) showed the nonexistence of Buonomano's memory effect because this effect was assumed to be a long range effect (in time) from the beginning.
- the Paris–Nord experiment[10] (with an atomic interferometer) showed that if the detectors of individual photons emitted by atoms at the outcome of the

atomic interferometer exhibit a randomization time, this time is certainly less than 1.5 microsec ($1.5 \ 10^{-6}$ s).

Despite of these negative results, one unexpected result was obtained[21]:

Indeed, we noticed[21,22] that a short range Shnoll effect[20] [as predicted by Shnoll's hypothesis (b)] was effectively present, weak but significant. Now, as this effect is washed out by Faraday shielding the QRNG, our conclusion is that Shnoll's effect (at least its short-range version and in the experimental situation considered by us) is an artefact due to electro-magnetic pollution of various sources (mobile phones, FM radio modulation and so on). It could be due nevertheless to by solar activity or by the cosmic background radiation as suggested by Shnoll's hypothesis (c). At this level, it is still too early to draw definitive conclusions from our analysis.

New experiments are welcome, among which the ones suggested by Sheldrake's[12] and Smolin's[11] models that we mentioned at the end of the previous section. In last resort, experiment ought to decide, as always in physics.

Acknowledgments

The author acknowledges support from N.T.U. concerning kind hospitality in Sgnpore and some help from F. Van den Berghe (VUB) for testing persistency of Paris–Nord time series. Sincere thanks too to Marco Bischof for drawing attention on Shnoll's work, some years ago.

References

1. A. Tonomura *et al.*, *Am. J. Phys.* **57**, 117 (1989).
2. http://en.wikipedia.org/wiki/Clustering-illusion
3. L. de Broglie, *Perspectives en Microphysique* (Champs, Flammarion, Paris, 1993).
4. L. Diosi, N. Gisin and W. T. Strunz, *Phys. Rev. A* **58**(3), 1699 (1998).
5. W. T. Strunz, L. Diosi and N. Gisin, *Phys. Rev. Lett.* **82**(9), 1801 (1999).
6. D. T. Gillespie, *Am. J. Phys.* **54**(10), 889 (1986).
7. D. T. Gillespie, *Phys. Rev. A* **49**(3), 1607 (1994).
8. G. Skorobogatov and R. Svertilov, *Phys. Rev. A* **58**(5), 3426 (1998).
9. J. Baudon *et al.*, *Comments At. Mol. Phys.* **34**(3–6), 161 (1999).
10. T. Durt *et al.*, Memory effects in atomic interferometry: A negative result in *Probing the Structure of Quantum Mechanics: Nonlinearity, Nonlocality, Computation and Axiomatics*, eds. D. Aerts, M. Czachor and T. Durt (World Scientific, Singapore, 2002), pp. 165–204.
11. L. Smolin, *A Real Ensemble Interpretation of Quantum Mechanics* (2011), arxiv: quant-ph/11042822.
12. R. Sheldrake, *Une Nouvelle Science de la vie*, ed. du Rocher (Monaco, 1985).
13. D. Bohm and J. Bub, *Rev. Mod. Phys.* **18**(1), 453 (1966).
14. C. Papaliolos, *Phys. Rev. Lett.* **18**(15), 622 (1967).
15. V. Buonomano and F. Bartmann, *Nuov. Cim. B* **95**(2), 99 (1986).
16. J. Summhammer, *Cim. B* **103**(3), 265 (1985).
17. V. Buonomano, *Founds. Phys. Lett.* **2**(6), 565 (1989).
18. D. Aerts, *J. Math. Phys.* **27**, 203 (1986).

19. T. Durt, From quantum to classical, a toy model, Doctoral Thesis, Vrije Universiteit Brussel (1996).
20. S. E. Shnoll *et al.*, *Physics-Uspekhi* **43**(2), 1025 (1998).
21. F. Van den Berghe, Quantum aspects of cryptography: from qutrit SIC POVM key encryption to randomness quality control, PhD, Vrije Universiteit Brussel (September 2011).
22. T. Durt, L.-P. Lamoureux and F. Van den Berghe, Shnoll effect in Vacuum fluctuations, in preparation.
23. L.-P. Lamoureux, SeQuR: Security by quantum randomness, a University of Brussels spin-off, 2007,
24. J. H. Tutsch, *Rev. Mod. Phys.* **40**, 232 (1968).
25. J. H. Tutsch, *Phys. Rev.* **183**(5), 1116 (1969).
26. F. J. Belinfante, *A Survey of Hidden Variable Theories* (Pergamon, Oxford, 1973).
27. J. S. Bell, *Physics* **1**, 195 (1964).
28. A. Aspect, P. Dalibard and G. Roger, *Phys. Rev. Lett.* **47**, 460 (1981).
29. T. Durt, *Helv. Phys. Acta* **72**, 356 (1999).
30. T. Durt, *Int. J. Theor. Phys.* **38**, 457 (1999).
31. R. Hanbury-Brown and R. Q. Twiss, *Proc. Roy. Soc. (London) A* **242**, 300 (1957).
32. R. Hanbury-Brown and R. Q. Twiss, *Proc. Roy. Soc. (London) A* **243**, 291 (1958).
33. L. Mandel and R. Short, *Phys. Rev. Lett.* **51**(5), 384 (1983).
34. L. Mandel and E. Wolf, *Optical Coherence and Quantum Optics* (Cambridge University Press, 1995).
35. B. Mandelbrot, *The Fractal Geometry of Nature*, eds. W. H. Freeman (New York, 1982).
36. T. Durt, L.-P. Lamoureux and F. Van den Berghe, *Characterization of a QRNG Based on Vacuum Fluctuations*, in preparation.
37. A. Einstein, B. Podolsky and N. Rosen, *Phys. Rev.* **47**, 777 (1935).
38. J. S. Bell, *Speakable and Unspeakable in Quantum Mechanics* (Cambridge University Press, 1987).
39. J. C. Vink, *Phys. Rev. A* **48**(3), 1808 (1993).
40. G. Cerofolini, *Lettere al Nuovo Cimento* **35**(15), 457 (1982).
41. H. Rauch, J. Summhammer and M. Zawisky, *Phys. Rev. A* **50**, 5000 (1994).
42. A. Valentini and H. Westman, *Dynamical Origin of Quantum Probabilities* (2004) arxiv: quant-ph/0403034.
43. Z. Merali, *Nature News* (2008). http://www.nature.com/news/2008/080515/full/news.2008.829.html
44. S. Colin and W. Struyve, Quantum nonequilibrium and relaxation to equilibrium for a class od de Broglie–Bohm-type theories", (2010) arxiv:quant-ph/0911.2823.
45. http://en.wikipedia.org/wiki/Gambler's-fallacy

Chapter 6

EXPLICIT FORMULA OF THE SEPARABILITY CRITERION FOR CONTINUOUS VARIABLES SYSTEMS

KAZUO FUJIKAWA

Institute of Quantum Science, College of Science and Technology,
Nihon University, Chiyoda-ku, Tokyo 101-8308, Japan

A very explicit analytic formula of the separability criterion of two-party Gaussian systems is given. This formula is compared to the past formulation of the separability criterion of continuous variables two-party Gaussian systems.

Keywords: Separability criterion; entanglement; continuous-variable two-party Gaussian system.

1. Introduction and Summary

The separability criterion of continuous variables systems is important not only for theoretical interest but also for practical applications in quantum information processing. The most basic example of continuous variables systems is the two-party Gaussian system, which may be compared to the most basic two-qubit system in the case of discrete variable models. This study of the separability criterion of two-party Gaussian systems was initiated by Duan *et al.*[1] and by Simon.[2] The analyses by these authors are however based on the rather abstract "existence proofs" and thus not easy to understand for the average workers in the field. Moreover, these two works are based on quite different formulations and their mutual relation is not obvious at all.

We here present an explicit analytic formula of the necessary and sufficient separability criterion of two-party Gaussian systems,[3,4] which should be useful to the wider audience in the field. We also clarify the difference in the above two approaches[1,2] explicitly.

To be specific, we show:

(i) We start with the 4×4 correlation matrix $V = (V_{\mu\nu})$ where

$$V_{\mu\nu} = \frac{1}{2}\langle \Delta\hat{\xi}_\mu \Delta\hat{\xi}_\nu + \Delta\hat{\xi}_\nu \Delta\hat{\xi}_\mu \rangle = \frac{1}{2}\langle \{\Delta\hat{\xi}_\mu, \Delta\hat{\xi}_\nu\} \rangle \tag{1}$$

with $\Delta\hat{\xi}_\mu = \hat{\xi}_\mu - \langle \hat{\xi}_\mu \rangle$ in term of the variables $(\hat{\xi}_\mu) = (\hat{q}_1, \hat{p}_1, \hat{q}_2, \hat{p}_2)$ for the two one-dimensional systems specified by canonical variables $(\hat{q}_1, \hat{p}_1)$ and $(\hat{q}_2, \hat{p}_2)$. We generally define $\langle \hat{O} \rangle = \text{Tr} \hat{\rho} \hat{O}$ by using the density matrix $\hat{\rho}$.

For a given standard form of covariance matrix (i.e., second moment of correlations)

$$
V_0 = \begin{pmatrix} a & 0 & c_1 & 0 \\ 0 & a & 0 & c_2 \\ c_1 & 0 & b & 0 \\ 0 & c_2 & 0 & b \end{pmatrix},
\tag{2}
$$

which is obtained from the general V by applying the $Sp(2, R) \otimes Sp(2, R)$ transformations,[2] the explicit form of *separability criterion* (which is in general a necessary condition) is given by:

$$
a \geq \frac{1}{2}, \quad b \geq \frac{1}{2},
$$
$$
0 \leq |c_1| \leq \frac{1}{2t} \{ [2ab(1 + t^2) + t] - 2\sqrt{D(a, b, t)} \}^{1/2},
\tag{3}
$$

where we defined:

$$
0 \leq t = \left| \frac{c_2}{c_1} \right| \leq 1
\tag{4}
$$

without loss of generality, and

$$
D(a, b, t) \equiv \sqrt{a^2 b^2 (1 - t^2)^2 + t(a + bt)(at + b)}.
\tag{5}
$$

(ii) For the covariance matrix obtained from the standard form V_0 by a squeezing $S^{-1} \in Sp(2, R) \otimes Sp(2, R)$ parameterized by r_1 and r_2,

$$
V = S^{-1} V_0 (S^{-1})^T
$$
$$
= \begin{pmatrix} ar_1 & 0 & c_1 \sqrt{r_1 r_2} & 0 \\ 0 & a/r_1 & 0 & c_2/\sqrt{r_1 r_2} \\ c_1 \sqrt{r_1 r_2} & 0 & br_2 & 0 \\ 0 & c_2/\sqrt{r_1 r_2} & 0 & b/r_2 \end{pmatrix},
\tag{6}
$$

the *optimal squeezing parameters* of P-representation condition

$$
V - \frac{1}{2} I \geq 0,
\tag{7}
$$

for which one can write the P-representation for the density matrix, are given by:

$$
r_1 = \frac{1}{at + b} \{ ab(1 - t^2) + \sqrt{D(a, b, t)} \},
$$
$$
r_2 = \frac{1}{a + bt} \{ ab(1 - t^2) + \sqrt{D(a, b, t)} \}
\tag{8}
$$

with the same $D(a, b, t)$ in (5). By using these parameters r_1 and r_2, one can write the the P-representation condition (7) as:

$$|c_1| \le \sqrt{(ar_1 - 1/2)(br_2 - 1/2)}/\sqrt{r_1 r_2}$$

$$= \sqrt{(a/r_1 - 1/2)(b/r_2 - 1/2)}/(t/\sqrt{r_1 r_2})$$

$$= \frac{1}{2t}\{[2ab(1 + t^2) + t] - 2\sqrt{D(a, b, t)}\}^{1/2} \tag{9}$$

which agrees with (3). The P-representation defines a separable density matrix by definition, as is shown in (17), and thus the separability criterion in (3) provides the *necessary and sufficient separability criterion.*[3]

2. Details of Analyses

We now discuss the details of the above analyses in connection with the past works on the separability criterion.

2.1. *Simon's criterion*

First of all, our formula (3) is a solution of the algebraic condition of Simon[2]

$$4(ab - c_1^2)(ab - c_2^2) \ge (a^2 + b^2) + 2|c_1 c_2| - \frac{1}{4} \tag{10}$$

written for the standard form of covariance matrix (2). It is clear that $c_1 = c_2 = 0$ in (2) defines a separable system, and thus we may convert (10) to a condition on c_1 and c_2. We solved the condition (10) by introducing an auxiliary parameter t with $0 \le t = |c_2/c_1| \le 1$ without loss of generality. It is however important to recognize the fact that the algebraic condition (10) allows the parameters in the range $c_1 = c_2 = \infty$ also, which is not allowed by our solution (3). To understand this discrepancy, one may go back to the separability criterion of Simon derived from Peres criterion[5]

$$\begin{pmatrix} A & C \\ C^T & B \end{pmatrix} + \frac{i}{2}\begin{pmatrix} J & 0 \\ 0 & \pm J \end{pmatrix} \ge 0 \tag{11}$$

where 4×4 covariance matrix V is written in terms of 2×2 submatrices

$$V = \begin{pmatrix} A & C \\ C^T & B \end{pmatrix}. \tag{12}$$

One can confirm that the algebraic condition (10) is given by:

$$\det\left[V_0 + \frac{i}{2}\begin{pmatrix} J & 0 \\ 0 & \pm J \end{pmatrix}\right] \ge 0, \tag{13}$$

namely, the algebraic condition (10) does not encode the full information of the condition (11) given by Peres criterion. Note that a positive determinant does not imply a positive matrix. The full contents of (11) are expressed by taking the expectation value of (11) in the form $v^\dagger M v$ by the four-component complex vectors $v = (d \pm ig, f \pm ih)$ with four real two-component vectors $d \sim h$ as:

$$d^T A d + f^T B f + 2d^T C f + g^T A g + h^T B h + 2g^T C h \geq |d^T J g| + |f^T J h| \qquad (14)$$

which is $Sp(2, R) \otimes Sp(2, R)$ invariant. The condition (14) holds for any real two-component vectors $d \sim h$.

If one imposes subsidiary conditions $g = J^T d$ and $h = \pm J^T f$ in (14), one obtains a *weaker condition*

$$d^T A d + f^T B f + 2d^T C f + d^T J A J^T d + f^T J B J^T f \pm 2d^T J C J^T f$$

$$\geq (d^T d + f^T f) \qquad (15)$$

which is no more $Sp(2, R) \otimes Sp(2, R)$ invariant. This condition (15) is easier to analyze and one obtains

$$\sqrt{(2a - 1)(2b - 1)} \geq |c_1| + |c_2| \qquad (16)$$

with $a \geq 1/2$ and $b \geq 1/2$ for the standard form of the covariance matrix in (2). The condition (16) clearly excludes the parameter range $c_1 = c_2 = \infty$ allowed by Simon's condition (10), and we recover our condition (3).

2.2. *Gaussian states and P-representation*

The P-representation of the density matrix

$$\hat{\rho} = \int d^2\alpha \int d^2\beta P(\alpha, \beta) |\alpha, \beta\rangle \langle \alpha, \beta| \qquad (17)$$

is defined in terms of coherent states and manifestly *separable*, $|\alpha, \beta\rangle = |\alpha\rangle |\beta\rangle$, and characterized by a 4×4 matrix P in terms of covariance matrix V

$$P(\alpha, \beta) = \frac{\sqrt{\det P}}{4\pi^2} \exp\left\{-\frac{1}{2}(\alpha_1, \alpha_2, \beta_1, \beta_2) P(\alpha_1, \alpha_2, \beta_1, \beta_2)^T\right\}, \qquad (18)$$

where

$$P^{-1} = V - \frac{1}{2}I \geq 0, \qquad (19)$$

if the P-representation exists.

The P-representation condition $V - (1/2)I \geq 0$ implies in our notation in (14)

$$d^T A d + f^T B f + 2d^T C f \geq \frac{1}{2}(d^T d + f^T f) \qquad (20)$$

and adding the expression with d and f replaced by g and h in (20), respectively, we recover the separability condition (14)

$$d^T A d + f^T B f + 2d^T C f + g^T A g + h^T B h + 2g^T C h$$

$$\geq \frac{1}{2}(d^T d + g^T g) + \frac{1}{2}(f^T f + h^T h)$$

$$\geq |d^T J g| + |f^T J h|. \qquad (21)$$

Namely, we have shown that *P-representation* $\Rightarrow$ *separability condition* as it should since the *P*-representation is separable.

The condition $V - (1/2)I \geq 0$ is not invariant under $S(r_1, r_2) \in Sp(2, R) \otimes Sp(2, R)$, and thus we consider the general covariance matrix in (6).

Squeezing parameters r_1 and r_2, which give the *boundary* of the condition $V - (1/2)I \geq 0$ for given V_0, is specified by[3]:

$$\left(a - \frac{1}{2r_1}\right)\left(b - \frac{1}{2r_2}\right) = \frac{1}{t^2}\left[\left(a - \frac{1}{2}r_1\right)\left(b - \frac{1}{2}r_2\right)\right] \tag{22}$$

and

$$\frac{(ar_1 - 1/2)}{(a/r_1 - 1/2)} = \frac{(br_2 - 1/2)}{(b/r_2 - 1/2)}. \tag{23}$$

These two equations are explicitly solved, and we obtain the analytic formulas of *optimal* squeezing parameters[3];

$$r_1 = \frac{1}{at + b}\{ab(1 - t^2) + \sqrt{D(a, b, t)}\},$$

$$r_2 = \frac{1}{a + bt}\{ab(1 - t^2) + \sqrt{D(a, b, t)}\} \tag{24}$$

and the *P*-representable (*separable Gaussian state*) condition $V - (1/2)I \geq 0$ gives

$$|c_1| \leq \sqrt{(ar_1 - 1/2)(br_2 - 1/2)}/\sqrt{r_1 r_2}$$

$$= \sqrt{(a/r_1 - 1/2)(b/r_2 - 1/2)}/(t/\sqrt{r_1 r_2})$$

$$= \frac{1}{2t}\{[2ab(1 + t^2) + t] - 2\sqrt{D(a, b, t)}\}^{1/2}. \tag{25}$$

Given any standard form of covariance matrix V_0, we can write the separable *P*-representation if $|c_1|$ satisfies the above condition (25) for any given $a \geq 1/2$, $b \geq 1/2$ and $1 \geq t \geq 0$ by using our formulas of squeezing parameters r_1 and r_2. This establishes that our criterion in (3) provides the necessary and sufficient condition of separable Gaussian states.

Note that the squeezing, which ensures the maximum domain for $|c_1|$ in (25), is achieved at $2a \geq r_1 \geq 1$, $2b \geq r_2 \geq 1$ to be consistent with the *P*-representation.

2.3. *Duan–Giedke–Chirac–Zoller criterion*

The weaker condition (15), which is no more $Sp(2, R) \otimes Sp(2, R)$ invariant, gives rise to the condition for the matrix (6)

$$\sqrt{(ar_1 - 1/2)(br_2 - 1/2)} + \sqrt{(a/r_1 - 1/2)(b/r_2 - 1/2)}$$

$$\geq \sqrt{r_1 r_2}|c_1| + \frac{|c_2|}{\sqrt{r_1 r_2}}, \tag{26}$$

which is in fact the original form of the separability criterion of Duan–Giedke–Chirac–Zoller[1] based on EPR-like operators. The condition (26) is based on the condition (15), which is weaker than Simon's condition (14), *cannot ensure the P-representation by itself*. DGCZ then supplement their weaker condition by imposing an extra condition

$$\sqrt{(ar_1 - 1/2)(br_2 - 1/2)} - \sqrt{r_1 r_2}|c_1|$$
$$= \sqrt{(a/r_1 - 1/2)(b/r_2 - 1/2)} - |c_2|/\sqrt{r_1 r_2}. \tag{27}$$

The solution of this extra constraint in the range $2a \geq r_1 \geq 1$, $2b \geq r_2 \geq 1$, if found, can ensure P-representation. But, *no proof* of this is given in DGCZ paper (only the existence in the interval $\infty \geq r_1 \geq 1$ is shown), and thus their original proof is incomplete in this sense. Their proof is however completed later from a different direction.[3]

It was also later recognized that the weaker separability criterion (26) is sufficient to ensure P-representation at the boundary of the P-representation condition. Namely, if one uses our formulas for the optimal values of squeezing parameters in (24), one can confirm that the relation (25) is equivalent to the weaker separability condition (26).[3] In this sense, the extra condition (27) in Ref. 1 is not required in the analysis of separability condition of two-party Gaussian systems.

2.4. *Hierarchy of separability criterions*

It is also shown[4] that we can derive a condition stronger than Simon's condition in a general context of two-party systems by an analysis of uncertainty relation and its variants. This stronger criterion however becomes equivalent to Simon's condition for the Gaussian system. Simon's condition in turn becomes equivalent to the weaker DGCZ criterion at the boundary of the P-representation. We thus have an interesting hierarchy of separability criteria for the continuous variables two party systems.

3. Discussion and Related References

We have presented a very explicit necessary and sufficient separability condition (3) of two-party Gaussian systems. We finally quote some Refs. 6–17 which were very helpful in the formulation of this explicit analytic formula.

References

1. L. M. Duan *et al.*, *Phys. Rev. Lett.* **84**, 2722 (2000).
2. R. Simon, *Phys. Rev. Lett.* **84**, 2726 (2000).
3. K. Fujikawa, *Phys. Rev. A* **79**, 032334 (2009).
4. K. Fujikawa, *Phys. Rev. A* **80**, 012315 (2009).
5. A. Peres, *Phys. Rev. Lett.* **77**, 1413 (1996).
6. S. Mancini and S. Severini, *Electron. Notes Theor. Comput. Sci.* **169**, 121 (2007).
7. E. Shchukin and W. Vogel, *Phys. Rev. Lett.* **95**, 230502 (2005).
8. A. Miranowicz and M. Piani, *Phys. Rev. Lett.* **97**, 058901 (2006).
9. R. F. Werner and M. M. Wolf, *Phys. Rev. Lett.* **86**, 3658 (2001).

10. G. Vidal and R. F. Werner, *Phys. Rev. A* **65**, 032314 (2002).
11. B. G. Englert and K. Wodkiewicz, *Phys. Rev. A* **65**, 054303 (2002).
12. G. Giedke *et al.*, *Phys. Rev. Lett.* **87**, 167904 (2001).
13. S. Mancini *et al.*, *Phys. Rev. Lett.* **88**, 120401 (2002).
14. J. Eisert, S. Scheel and M. B. Plenio, *Phys. Rev. Lett.* **89**, 137903 (2002).
15. M. M. Wolf, J. Eisert and M. B. Plenio, *Phys. Rev. Lett.* **90**, 047904 (2003).
16. M. G. Raymer *et al.*, *Phys. Rev. A* **67**, 052104 (2003).
17. V. Giovannetti *et al.*, *Phy. Rev. A* **67**, 022320 (2003).

Chapter 7

YANG–BAXTER EQUATIONS IN QUANTUM INFORMATION

MO-LIN GE

Theoretical Physics Section, Chern Institute of Mathematics,
Nankai University, Tianjin 300071, China
geml@nankai.edu.cn

KANG XUE

Department of Physics, Northeast Normal University,
Changchun, Ji Lin 120024, China

The connection between Yang–Baxter system and quantum information has been discussed. Based on the topological basis for both Temperley–Lieb (TL) algebra and Birman–Wenzl (BW) algebra the representations of N by N braiding matrices associated with the corresponding N^2 by N^2 ones are obtained. Some of physical consequences of the braiding matrices connecting with quantum information are shown.

Keywords: Yang–Baxter equation; topological basis; Birman–Wenzl algebra.

1. Introduction

The Yang–Baxter system plays important role in statistical models, many-body problems and quantum integrable models.[1–10] In the recent years the new type of braiding matrices and solutions of Yang–Baxter equation (YBE) has been found to be related to quantum information.[11–20] Furthermore, based on Yang–Baxterization[21–29] of the solutions of YBE associated with quantum information (type 2), the Berry's phase, Hamiltonian model and ℓ_1-norm property had been discussed.

In this talk we summarize the derived results related to TL and BW algebras[30–33] and make the extension. The interesting result in physics is that the velocity additivity rule for the type 2 solutions of YBE is no longer Galileo, but Lorentz type.

2. Two Types of Braiding Matrices. Yang–Baxter Equation and Temperley–Lieb Algebra

The YBE reads:

$$\check{R}_1(x)\check{R}_2(xy)\check{R}_1(y) = \check{R}_2(y)\check{R}_1(xy)\check{R}_2(x) \tag{1}$$

whose asymptotic one, i.e., braid matrix (BRM) obeys

$$B_1 B_2 B_1 = B_2 B_1 B_2, \tag{2}$$

where

$$
\begin{aligned}
B_1 &\equiv B_{12} = b(q) \otimes I, \\
B_2 &\equiv B_{23} = I \otimes b(q)
\end{aligned} \tag{3}
$$

and q is related to the coupling constant of the corresponding models. In general, the rational solution of (2) for the six-vortex models is called type-1 BRM. The typical one is:

$$
b' = \begin{vmatrix}
q & 0 & 0 & 0 \\
0 & 0 & -\eta & 0 \\
0 & -\eta^{-1} & q - q^{-1} & 0 \\
0 & 0 & 0 & \eta
\end{vmatrix} \tag{4}
$$

which is 4×4 representation of $b(q)$ and was known very well for years. Equation (4) can be regarded as q-deformation of the 4-D representation of permutation. For the later convenience we call b' type-1 BRM.

However, in the recent years the new development has been made to connect YBE with quantum information through the maximally entangled states, i.e., the Bell states defined by

$$
|\Phi^{\pm}\rangle = \frac{1}{\sqrt{2}}(|\uparrow\uparrow\rangle \pm |\downarrow\downarrow\rangle),
$$

$$
|\Psi^{\pm}\rangle = \frac{1}{\sqrt{2}}(|\uparrow\downarrow\rangle \pm |\downarrow\uparrow\rangle). \tag{5}
$$

The Bell states are connected to the natural basis $|\psi_0\rangle = (|\uparrow\uparrow\rangle, |\uparrow\downarrow\rangle, |\downarrow\uparrow\rangle, |\downarrow\downarrow\rangle)^T$ by a unitary transformation matrix W, which satisfies

$$(\Phi^+, \Psi^+, -\Psi^-, -\Phi^-) = W(|\uparrow\uparrow\rangle, |\uparrow\downarrow\rangle, |\downarrow\uparrow\rangle, |\downarrow\downarrow\rangle)^T, \tag{6}$$

where[11]

$$
W = \frac{1}{\sqrt{2}} \begin{pmatrix}
1 & 0 & 0 & 1 \\
0 & 1 & 1 & 0 \\
0 & -1 & 1 & 0 \\
-1 & 0 & 0 & 1
\end{pmatrix}. \tag{7}
$$

The W satisfies (7) and can be extended to matrix b as type-2 BRM given by:

$$b(q) = \frac{1}{\sqrt{2}} \begin{pmatrix} 1 & 0 & 0 & 1 \\ 0 & 1 & \epsilon & 0 \\ 0 & -\epsilon & 1 & 0 \\ -q^{-1} & 0 & 0 & 1 \end{pmatrix} = \frac{1}{\sqrt{2}}(1 + M), \qquad (8)$$

where

$$\epsilon^2 = 1, \quad M^2 = -1, \quad q = e^{i\alpha}.$$

Observing the two different types of braiding matrices b and b', both of them can be expressed in terms of matrix T as:

$$S = \rho(1 + fT), \qquad (9)$$

where S can be either b or b'. Constant f and matrix T can be defined through (4) or (8). For the type-1 we have:

$$T' = \begin{pmatrix} 1 & 0 & 0 & \eta \\ 0 & 0 & 0 & 0 \\ 0 & 0 & 0 & 0 \\ \eta^{-1} & 0 & 0 & 1 \end{pmatrix} \qquad (10)$$

whereas for the type-2 it leads to:

$$T = \frac{1}{\sqrt{2}} \begin{pmatrix} 1 & 0 & 0 & e^{i\alpha} \\ 0 & 1 & -i\epsilon & 0 \\ 0 & i\epsilon & 1 & 0 \\ e^{-i\alpha} & 0 & 0 & 1 \end{pmatrix}, \quad (\epsilon^2 = 1). \qquad (11)$$

Both of T and T' satisfy the Temperley–Lieb (T–L) algebra

$$T_i^2 = dT_i, \quad (T_i \equiv T_{i,i+1}) \qquad (12)$$

$$T_i T_{i+1} T_i = T_i, \quad (d = \text{loopvalue}), \qquad (13)$$

where d is constant given by a loop. The relation which satisfies (12) and (13) form the T–L algebra,[14] which originated in spin chain model. In general, the operators $\hat{T}_i$ work for any dimensional S. The point is to find the 2D representation of T' and T corresponding to the 4D representations of the operator $\hat{T}_i$.

Even both T' and T source from the different physical models describing integrable system and quantum information, respectively, but obey the same T–L algebra. There are two goals:

- find the 2D representation of T' and T, then make Yang–Baxterization[27–29];
- find the physical applications.

The graphic operator forms of $\hat{T}_i$ read

$$\hat{T}_i \equiv \hat{T}_{i,i+1} = \begin{array}{c}\sqcup\\\sqcap\end{array},$$

$$\hat{T}_i^2 = d\hat{T}_i = \begin{array}{c}\sqcup\\\bigcirc\\\sqcap\end{array} = \bigcirc \begin{array}{c}\sqcup\\\sqcap\end{array}, \quad d = \bigcirc \text{ (loop)}, \tag{14}$$

$$\hat{T}_i \hat{T}_{i+1} \hat{T}_i = \begin{array}{c}\sqcup\sqcup\\\ulcorner\sqcap\\\sqcap\end{array} = \begin{array}{c}\sqcup\\\sqcap\end{array}\Big|_{i\ i+1} = T_i.$$

Y–B operator can be written as:

$$\hat{S}(x) = \rho[I + G(x)\hat{T}]. \tag{15}$$

The value of a loop for T (type-2), $d = \sqrt{2}$, whereas for T' (type-1), $d = (q + q^{-1})$, i.e., $d = 2$ at $q = 1$. The braiding matrix can be expressed as:

$$\hat{S} = \rho\left(1 + f \begin{array}{c}\sqcup\\\sqcap\end{array}\right) = \times. \tag{16}$$

Following Kauffman,[34] there is the decomposition:

$$\check{S} = \times = \alpha\,\Big|\,\Big| + \alpha^{-1}\begin{array}{c}\sqcup\\\sqcap\end{array}, \tag{17}$$

$$d = -(\alpha^2 + \alpha^{-2}),$$

$$f = \frac{1}{2}\left(-d \pm \sqrt{d^2 - 4}\right).$$

In accordance with (10) and (11), for type-2 ($d = \sqrt{2}$), $f_2 = (-1)e^{\pm i\pi/4}$, while $f_1 = -1$ at $q = 1$ for type-1 ($d = 2$).

In terms of two fermions, the operators T_{ij} for T' and T take the forms:

$$\hat{T}'_{ij} = \frac{1}{\sqrt{2}}[I_{ij} + e^{i\alpha}S_i^+ S_j^+ + e^{-i\alpha}S_i^- S_j^- + i\epsilon(S_i^+ S_j^- - S_i^- S_j^+)],$$

$$\hat{T}_{ij} = \frac{1}{2}(I_{ij} + 4S_i^z S_j^z) + e^{i\alpha}S_i^+ S_j^+ + e^{-i\alpha}S_i^- S_j^-, \tag{18}$$

where i and j indicate the specified spaces and $\hat{T}_{ij}|k\rangle = |k\rangle$ for $k \neq i$, $k \neq j$, and S is spin operator.

By taking the elements $\langle\psi_0|\hat{T}'_{12}|\psi_0\rangle$ and $\langle\psi_0|\hat{T}_{12}|\psi_0\rangle$, the corresponding S-matrix (15) satisfying YBE for type-2 takes the form:

$$\check{R}'(\theta, \alpha) = \begin{pmatrix} \cos\theta & 0 & 0 & e^{i\alpha}\sin\theta \\ 0 & \cos\theta & \sin\theta & 0 \\ 0 & -\sin\theta & \cos\theta & 0 \\ -e^{-i\alpha}\sin\theta & 0 & 0 & \cos\theta \end{pmatrix}, \tag{19}$$

with $\cos\theta = (1-x)/\sqrt{2(1+x^2)}$ for type-2. For type-1, $x = e^{iu}$, the YBE is written in the familiar form in terms of u:

$$\check{R}'_1(u_1)\check{R}'_2(u_1+u_3)\check{R}'_1(u_3) = \check{R}'_2(u_3)\check{R}'_1(u_1+u_3)\check{R}'_2(u_1)\,, \tag{20}$$

$$\check{R}(u)\hat{S}(u) = \rho(u)[I + G(u)\hat{T}]\,, \tag{21}$$

whose 4D representation is:

$$\check{R}' = I + uP\,, \quad P(\eta) = \begin{pmatrix} 1 & 0 & 0 & 0 \\ 0 & 0 & \eta & 0 \\ 0 & -\eta^{-1} & 0 & 0 \\ 0 & 0 & 0 & 1 \end{pmatrix}\,, \tag{22}$$

For $\eta = -1$, it is the permutation.

3. 2-D Topological Basis of Braiding Matrices

The well-known set of basis is given by[13–16]

$$|e_1\rangle = \frac{1}{d} \underset{1\ 2}{\bigcup}\ \underset{3\ 4}{\bigcup}\,, \tag{23}$$

$$|e_2\rangle = \frac{\epsilon}{\sqrt{d^2-1}} \left(\underset{1\quad 2\quad 3\quad 4}{\bigsqcap} - \frac{1}{d}\underset{1\ 2}{\bigcup}\ \underset{3\ 4}{\bigcup} \right)\,, \tag{24}$$

where $\epsilon = \pm 1$, $|e_1\rangle$ and $|e_2\rangle$ are orthogonal basis and describe 2-component anyon.[17–20]

The braiding operations $\hat{A}$ and $\hat{B}$ act on $|e_1\rangle$ and $|e_2\rangle$ (Wilczek–Nayak, Freedman, Wang, ...)[16–20]:

$$\hat{A}: \quad \underset{1\ 2\ 3\ 4}{\bigslant} \quad \text{braiding the particles 1 and 2,} \tag{25}$$

$$\hat{B}: \quad \underset{1\ 2\ 3\ 4}{\bigslant} \quad \text{braiding the particles 2 and 3.} \tag{26}$$

$$\check{R} = \times = \alpha\ \Big\|\ + \alpha^{-1}\underset{\sqcap}{\sqcup}\,, \quad d = -(\alpha^2 + \alpha^{-2})\,. \tag{27}$$

For both of the types we act the operator $\hat{T}$ on $|e_1\rangle$ and $|e_2\rangle$ that lead to the dimensional representation of $\hat{T}$:

$$\hat{T}_{12}|e_1\rangle = \hat{T}_{34}|e_1\rangle = d|e_1\rangle\,, \quad \hat{T}_{12}|e_2\rangle = \hat{T}_{34}|e_2\rangle = 0\,, \tag{28}$$

$$\hat{T}_{23}|e_1\rangle = \hat{T}_{41}|e_1\rangle = \frac{1}{d}\left(|e_1\rangle + \epsilon\sqrt{d^2-1}|e_2\rangle\right)\,, \tag{29}$$

$$\hat{T}_{23}|e_2\rangle = \hat{T}_{41}|e_2\rangle = \frac{\sqrt{d^2-1}}{d}\left(\epsilon|e_1\rangle + \sqrt{d^2-1}|e_2\rangle\right)\,. \tag{30}$$

where the parameter d represents the value of a loop, i.e., $d = \sqrt{2}$ for type-2 and $d = 2$ for type-1 at $q = 1$, the matrices A and B take the explicit form:

$$A = \begin{pmatrix} (\alpha + \alpha^{-1})d & 0 \\ 0 & \alpha \end{pmatrix}, \quad \hat{A}\begin{pmatrix} |e_1\rangle \\ |e_2\rangle \end{pmatrix} = A\begin{pmatrix} |e_1\rangle \\ |e_2\rangle \end{pmatrix},$$

$$B = \frac{1}{\alpha d}\begin{pmatrix} (1 + \alpha^2 d) & \sqrt{d^2 - 1} \\ \sqrt{d^2 - 1} & \alpha^2 d + (d^2 - 1) \end{pmatrix}, \quad \hat{B}\begin{pmatrix} |e_1\rangle \\ |e_2\rangle \end{pmatrix} = B\begin{pmatrix} |e_1\rangle \\ |e_2\rangle \end{pmatrix}.$$

For $d = \sqrt{2}(\alpha = e^{i3\pi/8})$, we have for quantum information:

$$A = A_1 = e^{-i\pi/8}\begin{pmatrix} 1 & 0 \\ 0 & i \end{pmatrix}, \quad (ABA = BAB)$$

$$B = B_1 = \frac{1}{\sqrt{2}}e^{i\pi/8}\begin{pmatrix} 1 & -i \\ -i & 1 \end{pmatrix}.$$

(31)

For $d = 2$, $\alpha = i$, we obtain for spin chain models:

$$A_2 = \begin{pmatrix} -1 & 0 \\ 0 & 1 \end{pmatrix}, \quad B_2 = -\frac{1}{2}\begin{pmatrix} 1 & -\sqrt{3} \\ -\sqrt{3} & -1 \end{pmatrix}.$$

(32)

$$(A_2 B_2 A_2 = B_2 A_2 B_2),$$

where the overall factor i has been dropped.

The corresponding $\mathscr{A}(u)$ and $\mathscr{B}(u)$ satisfy YBE ($u = \tan\theta/2$) for the type-2:

$$\mathscr{A}_{\mathrm{I}}(u)\mathscr{B}_{\mathrm{I}}\left(\frac{u+v}{1+uv}\right)\mathscr{A}_{\mathrm{I}}(v) = \mathscr{B}_{\mathrm{I}}(v)\mathscr{A}_{\mathrm{I}}\left(\frac{u+v}{1+uv}\right)\mathscr{B}_{\mathrm{I}}(u). \quad \text{(Lorentz rule)}.$$

The velocity additivity for two anyons related to quantum information obeys the Lorentz form rather than the Galileo's ($c = 1$), if supposing their scattering obeys YBE. Since type-2 corresponds to anyonic picture with two-components, we expect that the velocity additivity rule of two anyons may not obey the Galileo formula.

For the type-2 by acting the operator $\hat{T}$ acts on $|e_1\rangle$ and $|e_2\rangle$ in terms of the usual spin basis at ith and jth spaces, we find

$$\hat{T}_{ij} = \sqrt{2}(|\psi_{ij}\rangle\langle\psi_{ij}| + |\phi_{ij}\rangle\langle\phi_{ij}|),$$

where

$$|\psi_{ij}\rangle = \frac{1}{\sqrt{2}}\left(|\underset{i}{\uparrow}\underset{j}{\uparrow}\rangle + e^{-i\alpha}|\underset{i}{\downarrow}\underset{j}{\downarrow}\rangle\right),$$

$$|\phi_{ij}\rangle = \frac{1}{\sqrt{2}}\left(|\underset{i}{\uparrow}\underset{j}{\downarrow}\rangle - i|\underset{i}{\downarrow}\underset{j}{\uparrow}\rangle\right).$$

(33)

Correspondingly,

$$|e_1\rangle = \frac{1}{\sqrt{2}}(|\psi_{12}\rangle|\psi_{34}\rangle + |\phi_{12}\rangle|\phi_{34}\rangle),$$

$$|e_2\rangle = \frac{1}{\sqrt{2}}\left[(1 - i\epsilon e^{i\alpha})|\psi_{23}\rangle|\psi_{41}\rangle - (1 - i\epsilon e^{i\alpha})|\phi_{23}\rangle|\phi_{41}\rangle - |e_1\rangle\right],$$

(34)

whereas for type-1, we have:

$$|e_1'\rangle = |\psi_{12}\rangle|\psi_{34}\rangle,$$

$$\underset{i\ j}{\overset{i\ j}{\sqcup}} = |\uparrow\uparrow\rangle + e^{-i\alpha}|\downarrow\downarrow\rangle = \sqrt{2}|\psi_{ij}\rangle, \quad (j = i + 1),$$

$$\underset{i\ j}{\sqcap} = \sqrt{2}\langle\psi_{ij}|,$$

$$\underset{i\ j}{\overset{\sqcup}{\sqcap}} = \hat{T}_i = 2|\psi_{ij}\rangle\langle\psi_{ij}|, \quad (j = i + 1).$$

4. Unified Form for Both Type-1 and Type-2 BRM's

Choosing the natural basis $|E_1\rangle$ and $|E_2\rangle$, they subject to the transformation:

$$\begin{pmatrix} |E_1\rangle \\ |E_2\rangle \end{pmatrix} = D^{1/2}(\theta,\varphi) \begin{pmatrix} |1\rangle \\ |2\rangle \end{pmatrix} = \begin{pmatrix} \cos\dfrac{\theta}{2} & -\sin\dfrac{\theta}{2}e^{-i\varphi} \\ \sin\dfrac{\theta}{2}e^{i\varphi} & \cos\dfrac{\theta}{2} \end{pmatrix} \begin{pmatrix} |1\rangle \\ |2\rangle \end{pmatrix}. \quad (35)$$

$D^{1/2}(\theta,\varphi)$ is the matrix form of Wigner's D-function with $J = 1/2$. $|1\rangle = |e_1\rangle$, $|2\rangle = |e_2\rangle$ or $|1\rangle = |\uparrow\uparrow \cdots \uparrow\rangle$, $|2\rangle = |\downarrow\downarrow \cdots \downarrow\rangle$, etc. The D-function $D(\theta,\varphi)$ means a rotation of angle θ about the axis **m**, which is determined by φ (**m** = $(-\sin\varphi, \cos\varphi, 0)$).

To satisfy the 2-D braid relation:

$$D(\theta,\varphi_1)D(\theta,\varphi_2)D(\theta,\varphi_1) = D(\theta,\varphi_2)D(\theta,\varphi_1)D(\theta,\varphi_2), \quad (36)$$

then θ and φ should obey the relation[35]

$$\cos\varphi = \frac{\cos\theta}{1 - \cos\theta}, \quad (37)$$

where $\varphi = \varphi_2 - \varphi_1$. Setting $\varphi_1 = 0$ and $\varphi_2 = \varphi$ for simplicity, we get

$$A(\theta) = D(\theta,\varphi_1 = 0) = \begin{pmatrix} \cos\dfrac{\theta}{2} & -\sin\dfrac{\theta}{2} \\ \sin\dfrac{\theta}{2} & \cos\dfrac{\theta}{2} \end{pmatrix}, \quad (38)$$

$$B(\theta) = D(\theta, \varphi_2 = \varphi) = \begin{pmatrix} \cos\dfrac{\theta}{2} & -\sin\dfrac{\theta}{2}e^{-i\varphi} \\ \sin\dfrac{\theta}{2}e^{i\varphi} & \cos\dfrac{\theta}{2} \end{pmatrix}. \tag{39}$$

Clearly $A(\theta)$ and $B(\theta)$ satisfy braid relation for arbitrary θ:

$$A(\theta)B(\theta)A(\theta) = B(\theta)A(\theta)B(\theta). \tag{40}$$

It is emphasized that two different φ's specify $A(\theta)$ and $B(\theta)$ satisfying (40). Making the unitary transformation:

$$V A(\theta) V^{\dagger} = \begin{pmatrix} e^{i\theta/2} & 0 \\ 0 & e^{-i\theta/2} \end{pmatrix}, \tag{41}$$

and

$$V B(\theta) V^{\dagger} = \begin{pmatrix} \cos\dfrac{\theta}{2} + i\sin\dfrac{\theta}{2}\cos\varphi & i\sin\varphi\sin\dfrac{\theta}{2} \\ i\sin\varphi\sin\dfrac{\theta}{2} & \cos\dfrac{\theta}{2} - i\sin\dfrac{\theta}{2}\cos\varphi \end{pmatrix}, \tag{42}$$

where $V = (1/\sqrt{2})\begin{pmatrix} 1 & i \\ i & 1 \end{pmatrix}$ the matrix A is diagonalized.

(i) $\varphi = \pi/2$, $\theta = -\pi/2$, Eqs. (41) and (42) become into $e^{-i\pi/4}\begin{pmatrix} 1 & 0 \\ 0 & i \end{pmatrix}$ and $(1/\sqrt{2})\begin{pmatrix} 1 & -i \\ -i & 1 \end{pmatrix}$ respectively. By adjusting the phase factor, we re-obtain the anyonic description.

(ii) $\varphi = 2\pi/3$, $\theta = \pi$, Eqs. (41) and (42) become into $(-i)\begin{pmatrix} -1 & 0 \\ 0 & 1 \end{pmatrix}$ and $(-(i/2))\begin{pmatrix} 1 & -\sqrt{3} \\ -\sqrt{3} & -1 \end{pmatrix}$, that is 2-D representation of braiding matrix for the usual spin chain model.

Correspondingly, for type-2, the 4×4 $\breve{R}$-matrix is found to be:

$$\breve{R}(\theta, \varphi) = \begin{pmatrix} \cos\theta & 0 & 0 & e^{-i\varphi}\sin\theta \\ 0 & \cos\theta & \sin\theta & 0 \\ 0 & -\sin\theta & \cos\theta & 0 \\ -e^{i\varphi}\sin\theta & 0 & 0 & \cos\theta \end{pmatrix}.$$

For the anyonic picture of quantum information, $\theta = -\pi/2$, i.e., $\varphi = \pi/2$,

$$A_1 = \frac{1}{\sqrt{2}}\begin{pmatrix} 1 & 1 \\ -1 & 1 \end{pmatrix}, \quad B_1 = \frac{1}{\sqrt{2}}\begin{pmatrix} 1 & -i \\ -i & 1 \end{pmatrix}.$$

Using the unitary transformation $V = (1/\sqrt{2})\begin{pmatrix} 1 & i \\ i & 1 \end{pmatrix}$ to make matrix A_1 diagonal:

$$A'_1 = V A_1 V^\dagger = e^{-i\pi/4} \begin{pmatrix} 1 & 0 \\ 0 & i \end{pmatrix}, \tag{43}$$

$$B'_1 = V B_1 V^\dagger = \frac{1}{\sqrt{2}} \begin{pmatrix} 1 & -i \\ -i & 1 \end{pmatrix}. \tag{44}$$

The 2-D YBE then has the form:

$$A(\theta_1)B\left(\theta_2, \varphi = \frac{\pi}{2}\right)A(\theta_3) = B\left(\theta_3, \varphi = \frac{\pi}{2}\right)A(\theta_2)B\left(\theta_1, \varphi = \frac{\pi}{2}\right), \tag{45}$$

where the spectral parameters should satisfy for anyons:

$$\tan \frac{\theta_2}{2} = \frac{\tan \dfrac{\theta_1}{2} + \tan \dfrac{\theta_3}{2}}{1 + \tan \dfrac{\theta_1}{2} \tan \dfrac{\theta_3}{2}}. \tag{46}$$

By setting $\tan(\theta/2) = \beta u$, this is just the additivity rule of Lorentz velocity:

$$u_2 = \frac{u_1 + u_3}{1 + \beta^2 u_1 u_3}.$$

The discussion on the value of β is dropped.

For the type-1 BRM (for spin models), $\theta = \pi$, i.e., $\varphi = 2\pi/3$,

$$A_2 = \frac{1}{\sqrt{2}} \begin{pmatrix} 0 & -1 \\ 1 & 0 \end{pmatrix}, \quad B_2 = \frac{1}{\sqrt{2}} \begin{pmatrix} 0 & -e^{-2i\pi/3} \\ e^{2i\pi/3} & 0 \end{pmatrix}.$$

Taking the same unitary transformation as for type-2, we have:

$$A'_2 = V A_2 V^\dagger = (-i) \begin{pmatrix} -1 & 0 \\ 0 & 1 \end{pmatrix}, \tag{47}$$

$$B'_2 = V B_2 V^\dagger = \left(-\frac{i}{2}\right) \begin{pmatrix} 1 & -\sqrt{3} \\ -\sqrt{3} & -1 \end{pmatrix}. \tag{48}$$

The corresponding 2-D YBE reads:

$$A(\theta_1)B\left(\theta_2, \varphi = \frac{2\pi}{3}\right)A(\theta_3) = B\left(\theta_3, \varphi = \frac{2\pi}{3}\right)A(\theta_2)B\left(\theta_1, \varphi = \frac{2\pi}{3}\right), \tag{49}$$

The spectral parameters should satisfy the Galileo relation for:

$$u = \tan \frac{\theta}{2}, \tag{50}$$

$$u_2 = u_1 + u_3.$$

That has been known well.

5. Two Types of BRM in Physics and Extremization of ℓ_1-Norm

The extremization of ℓ_1-norm of the Wigner's D-function for spin $1/2$:

- maximizing $\sum_{M=-1/2}^{1/2} |D_{MM'}^{1/2}(\theta, \varphi)|$ to yield type-1 BRM;
- maximizing $\sum_{M=-1/2}^{1/2} |D_{MM'}^{1/2}(\theta, \varphi)|$ to yield type-2 BRM.

In principle, can be extended to any dimensional spin representations.[26]

5.1. *Motivation of ℓ_1-norm involution*

If a signal is sparse, then much less measurements y may be made to recover $f(t) = x$. Suppose measuring matrix of Φ is of $M \times N$ matrix ($M \approx k \log N \ll N$, where k is "sparsity"), i.e., $y = \Phi x$. To recover x (N components), we can only measure M data. Obviously, for given y to find x is an ill-posted problem because it does not have the inverse. However, the Compressive Sensing (Donoho–Candes–Tao) tells that the recovery of $f(t) = x$ consists in Refs. 36–38.

$$\text{minimize} \quad \|x\|_{\ell_1} \quad \text{subject to} \quad y = \Phi x. \tag{51}$$

The ℓ_1-norm plays the crucial role in (51). Through this example, we learn that the minimization of ℓ_1-norm can be used to determine some important physical quantities.

5.2. *Extremization of D-function and ℓ_1-norm*

We should emphasize that only two sets of θ and φ, i.e., $\{\theta = -\pi/2, \varphi = \pi/2\}$ and $\{\theta = \pi, \varphi = 2\pi/3\}$ have the "real" physical meanings. For the 4×4 form, there are only two types of matrices in physics, i.e., $\hat{T}_2$ and $\hat{T}_1$ are allowed that correspond to the familiar six-vertex model and quantum information (Bell states), respectively.

If we take the ℓ_1-norm of the coefficients of the decomposition of $|E_1\rangle$ and $|E_2\rangle$ in (35), we then have $f(\theta) = |\cos(\theta/2)| + |-\sin(\theta/2)e^{-i\varphi}|$.

The two basis satisfy the same relation for $J_z = \pm 1/2$. If θ is restricted in the field $[-\pi, \pi]$, then $\cos(\theta/2)$ is always positive. Also $\sin(\theta/2) \geqslant 0$ if $0 \leqslant \theta \leqslant \pi$, and $\sin(\theta/2) < 0$ if $-\pi \leqslant \theta < 0$. Using these results, we can easily calculate the maximum and minimum values of $f(\theta)$ and the corresponding θ. When $\theta \in [0, \pi]$, $f(\theta) = \cos(\theta/2) + \sin(\theta/2)$, $f(\theta)$ takes maximum value when $\theta = \pi/2$ while it takes minimum value when $\theta = 0, \pi$. When $\theta \in [-\pi, 0]$, $f(\theta) = \cos(\theta/2) - \sin(\theta/2)$, $f(\theta)$ takes maximum value when $\theta = -\pi/2$ and takes minimum value when $\theta = -\pi$.[26]

6. Bogoliubov's Hamiltonian Induced from Dirac's Hamiltonian via Braid Matrix

As has been shown in the above the 4×4 realization of the $T = U$ operator for T–L algebra can be taken as

$$U = \frac{1}{\sqrt{2}} \begin{bmatrix} 1 & 0 & 0 & i\epsilon \\ 0 & 1 & i\epsilon & 0 \\ 0 & -i\epsilon & 1 & 0 \\ -i\epsilon & 0 & 0 & 1 \end{bmatrix}, \quad \epsilon^2 = 1. \tag{52}$$

In general:

$$U_{ii+1} = \frac{1}{\sqrt{2}} (I_{ii+1} - \epsilon \sigma_i^y \sigma_{i+1}^x) \tag{53}$$

where σ is Pauli matrix. The Dirac's Hamiltonian reads

$$H_D = mc^2 \beta + c\alpha \cdot \mathbf{p} \tag{54}$$

$$\beta = \begin{bmatrix} I & 0 \\ 0 & -I \end{bmatrix} \quad \alpha = \begin{bmatrix} 0 & \sigma \\ \sigma & 0 \end{bmatrix}.$$

As was pointed out that two matrices $A \sim \begin{bmatrix} 1 & 0 \\ 0 & i \end{bmatrix}$ and $B \sim \begin{bmatrix} 1 & i \\ i & 0 \end{bmatrix}$ obey braid relation $ABA = BAB$ by acting on the specified states. The Dirac Hamiltonian can be written as:

$$H_D = \frac{1}{E}(m\beta + \boldsymbol{\sigma} \cdot \boldsymbol{\alpha}) = \frac{1}{E}(m\sigma_3 \otimes \bullet + \sigma_1 \otimes \mathbf{p} \cdot \boldsymbol{\sigma}), \tag{55}$$

where the Pauli matrices $\sigma_1 = \begin{bmatrix} 0 & 1 \\ 1 & 0 \end{bmatrix}$, $\sigma_2 = \begin{bmatrix} 0 & -i \\ i & 0 \end{bmatrix}$ and $\sigma_3 = \begin{bmatrix} 1 & 0 \\ 0 & -1 \end{bmatrix}$ have been used and the Hamiltonian has been divided by energy E to normalize the eigenvalue of H_D to be unity.

In terms of the unitary transformation the Hamiltonian can be diagonalized as:

$$H_D' = V H_D V^\dagger = \begin{bmatrix} \sigma_3 & 0 \\ 0 & \sigma_3 \end{bmatrix} = \begin{bmatrix} H_{1D}' & 0 \\ 0 & H_{2D}' \end{bmatrix} \tag{56}$$

where V is given by:

$$V = \frac{1}{\sqrt{2E}} \begin{bmatrix} p_+ \mu_-^{-1} & -p_3 \mu_-^{-1} & 0 & \mu_- \\ -p_+ \mu_+^{-1} & p_3 \mu_+^{-1} & 0 & \mu_+ \\ p_3 \mu_-^{-1} & p_- \mu_-^{-1} & \mu_- & 0 \\ -p_3 \mu_+^{-1} & -p_- \mu_+^{-1} & \mu_+ & 0 \end{bmatrix} \tag{57}$$

where $p_\pm = p_x \pm i p_y$ and $\mu_\pm = (E \pm m)^{1/2}$. It is easy to show

$$V.V^\dagger = \bullet. \tag{58}$$

Under the transformation the Dirac Hamiltonian becomes block-diagonal. We introduce the matrix

$$A(\theta) = e^{i\frac{\phi}{2}H'_D} = e^{i\frac{\phi}{2}VH_DV^\dagger} = Ve^{i\frac{\phi}{2}H_D}V^\dagger = \begin{bmatrix} a(\phi) & 0 \\ 0 & a(\phi) \end{bmatrix} \tag{59}$$

where $a(\phi) = \begin{bmatrix} e^{i\phi/2} & 0 \\ 0 & e^{-i\phi/2} \end{bmatrix} = e^{i\phi/2}\begin{bmatrix} 1 & 0 \\ 0 & e^{-i\phi} \end{bmatrix}$.

If $\phi = -\pi/2$ (see Sec. 4)

$$a = a\left(\phi = -\frac{\pi}{2}\right) = e^{-i\frac{\pi}{4}}\begin{bmatrix} 1 & 0 \\ 0 & i \end{bmatrix} \tag{60}$$

then,

$$A = Ve^{-i\pi/4H_D}V^\dagger = \begin{bmatrix} 1 & 0 & 0 & 0 \\ 0 & i & 0 & 0 \\ 0 & 0 & 1 & 0 \\ 0 & 0 & 0 & i \end{bmatrix} \tag{61}$$

Following ABA=BAB, we get,

$$B = \frac{1}{\sqrt{2}}\begin{bmatrix} 1 & i & 0 & 0 \\ i & 1 & 0 & 0 \\ 0 & 0 & 1 & i \\ 0 & 0 & i & 1 \end{bmatrix}. \tag{62}$$

and define the new Hamiltonian $\mathcal{H}$ through the inverse transformation

$$Ve^{-i\frac{\pi}{4}\mathcal{H}}V^\dagger = B$$

$$\exp\left(-i\frac{\pi}{4}\mathcal{H}\right) = \sqrt{2}\left(\cos\frac{\pi}{4} - i\mathcal{H}\sin\frac{\pi}{4}\right) = 1 - i\mathcal{H} \tag{63}$$

$$\mathcal{H} = \frac{1}{E}\left(p\beta - \frac{m}{p}\mathbf{p}\cdot\boldsymbol{\alpha}\right), \quad E = (m^2 + p^2)^{1/2}, \tag{64}$$

where $p = |\mathbf{p}|$ and it is easy to check that $\mathcal{H}^2 = \bullet$.

The obtained Hamiltonian $\mathcal{H}$ can be recast to:

$$\mathcal{H} = \left(\frac{m}{p}\right)\frac{1}{E}\left(\frac{p^2}{m}\beta - \mathbf{p}\cdot\boldsymbol{\alpha}\right), \tag{65}$$

where

$$\mathcal{E} = \frac{p}{m}E. \tag{66}$$

The $\mathcal{H}$ normalized by $\mathcal{E}$ has the same form as Bogoliubov Hamiltonian for quasi particles in ^{3}He–B with the free energy $\mu = 0$ and mass being $(1/2)m$.

7. Optical Simulation of YBE

As was shown that there is equivalence between 4×4 $\check{R}(\theta, \phi)$-matrix satisfying the usual YBE and 2×2 $A(u)$ and $B(u)$ satisfying $ABA = BAB$ type of braid relation. The $A(u)$ and $B(u)$ matrices given in Sec. 7 can be recast to[21-25]

$$A(k) = a(k) \begin{bmatrix} \dfrac{k+ic}{k-ic} & 0 \\ 0 & 1 \end{bmatrix}, \quad B(k) = \dfrac{a(k)}{k-ic} \begin{bmatrix} k & ic \\ ic & k \end{bmatrix}, \tag{67}$$

where $1 + \beta^2 u^2 = k$ and $2\epsilon_\beta u = c(k)$. Using

$$\frac{k}{\sqrt{k^2 + c^2}} = \cos\frac{\theta}{2}, \quad \frac{c}{\sqrt{k^2 + c^2}} = \sin\frac{\theta}{2}. \tag{68}$$

Then that is equivalent to the YBE for $\check{R}$-matrix

$$A(k) = \alpha(k) \begin{bmatrix} 1 & 0 \\ 0 & e^{-i\frac{\theta}{2}} \end{bmatrix}$$

$$B(k) = \alpha(k) \begin{bmatrix} \cos\dfrac{\theta}{2} & \sin\dfrac{\theta}{2} e^{i\pi/2} \\ \sin\dfrac{\theta}{2} e^{i\pi/2} & \cos\dfrac{\theta}{2} \end{bmatrix}, \tag{69}$$

where $\alpha(k) = a(k)e^{i\theta(k)}$. Noting $u = (1/\beta)\sqrt{k-1}$, the matrices $A(u)$ and $B(u)$ satisfy

$$\boxed{A(u)B\left(\frac{u+v}{1+\beta^2 uv}\right)A(v) = B(v)A\left(\frac{u+v}{1+\beta^2 uv}\right)B(u)}$$

that is equivalent to the relation for the 4-D YBE for $\check{R}$-matrix.

We would like to show that the YBE can be checked in terms of optical scheme, i.e., YBE can be directly checked with the help of the usually optical experiment for chiral photons. Noting the convention

$$|0\rangle = \searrow, |1\rangle = \nearrow$$

and recalling Mach–Zehnder interferometer in which the two beam splitters have the ratio of transmission over reflection coefficients as $50\% = 50\%$, the following design for two chiral photon beams is exactly the operation of $B(k)$:

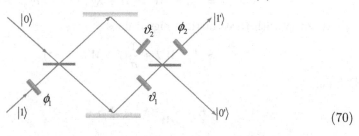

$$\tag{70}$$

where ϕ and θ represent phase shifts. Noting $\theta = \theta_2 - \theta_1$ and total phase shift $\Phi = \phi_1 + \phi_2 + \theta_1 + \theta_2$ the optical scheme gives the unitary operation:

$$V_{MZ} = e^{i\Phi/2} \begin{pmatrix} \cos(\theta/2)e^{-i(\phi_1+\phi_2)/2} & \sin(\theta/2)e^{i(\phi_1-\phi_2)/2} \\ -\sin(\theta/2)e^{-i(\phi_1-\phi_2)/2} & \cos(\theta/2)e^{i(\phi_1+\phi_2)/2} \end{pmatrix}.$$

Setting $\phi_1 = -\phi_2 = \pi/2$ and $\theta = \theta_2 - \theta_1$ arbitrary, Eq. (71) reduces to the $B(k)$ matrix given by Eq. (69). For simplicity we take $\theta_2 = -\theta_1 = \theta$ then $\Phi = 0$. Regardless the common factor $\alpha(k)$ the matrix $B(k)$ can be realized by V_{MZ} and the matrix $A(k)$ can easily be realized by phase shift operation. The YBE for A and B can then be checked through the following optical scheme.

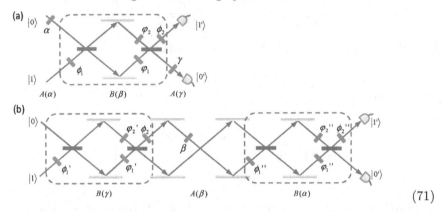

(71)

In terms of continuous adjustment of corresponding phases and identification of $A(\alpha)B(\beta)A(\gamma)$ and $B(\gamma)A(\beta)B(\alpha)$, the optical experiment of chiral photon (say, vertical and horizontal polarization) can then be used to check the YBE directly. It is interesting to check this scheme experimentally.

8. Topological Basis for BWA

The Birman–Wenzl algebra (BWA) relations are formed besides $\check{R}_{i,j+1} = S_{i,j+1}$, there is new operator $E_{i,j+1}$. Denoting the eigenvalues of a braiding matrix S with three eigenvalues by λ_1, λ_2 and λ_3, where S satisfies braid relation

$$S_{12}S_{23}S_{12} = S_{23}S_{12}S_{23} \quad (S_1 \equiv S_{12} = S \otimes I, S_2 = S_{23} = I \otimes S) \tag{72}$$

and without loss of generality by setting

$$\lambda_1\lambda_2 = -1, \quad W = \lambda_1 + \lambda_2, \quad \lambda_3 = \sigma \tag{73}$$

we have for S with three distinct eigenvalues

$$S - S^{-1} = WI + \frac{1}{\sigma}(I + WS - S^2). \tag{74}$$

Defining

$$E = \frac{1}{\sigma W}(S^2 - WS - I) \tag{75}$$

Eq. (74) becomes

$$S - S^{-1} = W(I - E).$$ (76)

They form BWA. Noting that a loop takes the value

$$d = 1 + \frac{1}{W}(\sigma^{-1} - \sigma)$$ (77)

and extending the topological basis $|e_1\rangle$ and $|e_2\rangle$ for T–L algebra, we shall find the uni-orthogonal basis $|e_1\rangle$, $|e_2\rangle$ and $|e_3\rangle$ such that:

$$S_{12}|e_\mu\rangle = \lambda_\mu|e_\mu\rangle \quad (\mu = 1, 2, 3)$$ (78)

with

$$S_{12} = \times \quad \text{and} \quad E_i = \underset{i \quad i+1}{\sqcap}, \quad d = \bigcirc \quad (i = 1, 2),$$ (79)

where the eigenvalues λ_μ may be complex.

The proposed topological basis for BWA is:

$$|e_3\rangle = d^{-1}\sqcup\sqcup$$ (80)

$$|e_i\rangle = f_i\left\{ \Big\rangle\!\!\!\times + \alpha_i \sqcup\!\sqcup + \beta_i \sqcup\sqcup \right\} \quad (i = 1, 2).$$ (81)

We shall prove that the (78) together with

$$\langle e_3|e_i\rangle = 0, \quad \langle e_i|e_j\rangle = \delta_{ij} \quad (i, j = 1, 2),$$ (82)

lead to the constraints to the parameters α_i and β_i and normalization constant f_i:

$$\alpha_i = \lambda_i \quad (\lambda_1\lambda_2 = -1), \quad \alpha_i + \beta_i d = -\sigma^{-1} \quad (i = 1, 2)$$ (83)

and

$$f_i = \{d(\lambda_i^2 + 1)[-\lambda_i^{-1}d^{-1}(\sigma^{-1} + \lambda_i) + \lambda_i^{-1}\sigma + d]\}^{-1/2}$$ (84)

for $\lambda_\mu^* = \lambda_\mu(\mu = 1, 2, 3)$, i.e., $S^\dagger = S$ (hermitian), whereas

$$f_i = \{(d - 1)(\lambda_i + \lambda_i^{-1})(\sigma + \lambda_i d + \lambda_i^{-1})\}^{-1/2}$$ (85)

for $\lambda_\mu^* = \lambda_\mu^{-1}$ ($\mu = 1, 2, 3$), i.e., $S^\dagger = S^{-1}$ (unitary). The (83) takes the same form for S being hermitian or unitary. The only difference between hermitian and unitary consists in the different f_i.

8.1. *Matrix form of E_{12}, E_{23}, S_{12} and S_{23} for $S^\dagger = S$*

In terms of the uni-orthogonal basis the direct calculation gives the matrix forms of E and S.

$$
\begin{aligned}
(E_A)_{\mu\nu} &= \langle e_\mu | E_{12} | e_\nu \rangle \\
(E_B)_{\mu\nu} &= \langle e_\mu | E_{23} | e_\nu \rangle \quad (\mu, \nu = 1, 2, 3) \\
A_{\mu\nu} &= \langle e_\mu | S_{12} | e_\nu \rangle \\
B_{\mu\nu} &= \langle e_\mu | S_{23} | e_\nu \rangle
\end{aligned}
\tag{86}
$$

The explicit 3-D matrix forms are given by ($\lambda_1 \lambda_2 = -1$):

$$
E_A = \begin{bmatrix} 0 \\ & 0 \\ & & d \end{bmatrix}, \quad
A = \begin{bmatrix} \lambda_1 \\ & \lambda_2 \\ & & \lambda_3 \end{bmatrix} = \begin{bmatrix} \lambda_1 \\ & -\lambda_1^{-1} \\ & & \sigma \end{bmatrix}
\tag{87}
$$

$$
E_B = d^{-1}(\lambda_1 + \lambda_1^{-1})^{-1}
$$
$$
\times \begin{bmatrix}
(\lambda_1 + \lambda_1^{-1})^{-1} f_1^{-2} & -(\lambda_1 + \lambda_1^{-1})^{-1} (f_1 f_2)^{-1} & f_1^{-1} \\
-(\lambda_1 + \lambda_1^{-1})^{-1} (f_1 f_2)^{-1} & (\lambda_1 + \lambda_1^{-1})^{-1} f_2^{-2} & -f_2^{-1} \\
f_1^{-1} & -f_2^{-1} & \lambda_1 + \lambda_1^{-1}
\end{bmatrix}
\tag{88}
$$

$$
B = d^{-1}(\lambda_1 + \lambda_1^{-1})^{-1}
$$
$$
\times \begin{bmatrix}
\lambda_1^{-1}(\lambda_3(d-1) - \lambda_1^{-1}) & (1 - \lambda_1^{-1}\lambda_3 - \lambda_1\lambda_3 d) f_1^{-1} f_2 & \lambda_1^{-1} f_1^{-1} \\
(1 - \lambda_1^{-1}\lambda_3 - \lambda_1\lambda_3 d) f_1^{-1} f_2 & -\lambda_1(\lambda_3(d-1) + \lambda_1) & \lambda_1 f_2^{-1} \\
\lambda_1^{-1} f_1^{-1} & \lambda_1 f_2^{-1} & (\lambda_1 + \lambda_1^{-1})\lambda_3^{-1}
\end{bmatrix}.
\tag{89}
$$

When $\lambda_1 = q$, $\lambda_2 = -q^{-1}$, $\lambda_3 = q^{-2}$,

$$
E_B = \begin{bmatrix}
d^{-1}(d^2 - d - 1) & -d^{-\frac{1}{2}}(d^2 - d - 1)^{\frac{1}{2}} & d^{-1}(d^2 - d - 1)^{\frac{1}{2}} \\
-d^{-\frac{1}{2}}(d^2 - d - 1)^{\frac{1}{2}} & 1 & -d^{-\frac{1}{2}} \\
d^{-1}(d^2 - d - 1)^{\frac{1}{2}} & -d^{-\frac{1}{2}} & d^{-1}
\end{bmatrix}
\tag{90}
$$

$$
B = \begin{bmatrix}
q^{-4} d^{-1}(d-1)^{-1} & -q^{-2} d^{-\frac{1}{2}}(d-1)^{-1}(d^2 - d - 1)^{\frac{1}{2}} & d^{-1} q^{-1}(d^2 - d - 1)^{\frac{1}{2}} \\
-q^{-2} d^{-\frac{1}{2}}(d-1)^{-1}(d^2 - d - 1)^{\frac{1}{2}} & (d-1)^{-1}(d-2) & d^{-\frac{1}{2}} q \\
d^{-1} q^{-1}(d^2 - d - 1)^{\frac{1}{2}} & d^{-\frac{1}{2}} q & d^{-1} q^2
\end{bmatrix}
\tag{91}
$$

$$
ABA = BAB
\tag{92}
$$

In case of $S^\dagger = S^{-1}$ the similar relations can also be set up.

8.2. *Application to the "standard" q-deformed spin-1 model*

Suppose

$$S = \text{Block diag}(A_1, A_2, A_3, A_2, A_1), \quad \omega = q - q^{-1} = W,$$

$$(\lambda_1 = q, \lambda_2 = -q^{-1}, \sigma = q^{-2}). \tag{93}$$

The "standard" generic S-matrix for spin-1 was given by[1,2,39,40]:

$$A_1 = q, \quad A_2 = \begin{bmatrix} 0 & 1 \\ 1 & q - q^{-1} \end{bmatrix} \tag{94}$$

$$A_3 = \begin{bmatrix} 0 & 0 & q^{-1} \\ 0 & 1 & -\omega q^{-\frac{1}{2}} \\ q^{-1} & -\omega q^{-\frac{1}{2}} & \omega(1 - q^{-1}) \end{bmatrix} \quad (q = \text{real}). \tag{95}$$

We find for the q-deformed spin-1 model:

$$E_3 = \begin{bmatrix} q & q^{\frac{1}{2}} & 1 \\ q^{\frac{1}{2}} & 1 & q^{-\frac{1}{2}} \\ 1 & q^{-\frac{1}{2}} & q^{-1} \end{bmatrix} \quad E_1 = E_2 = 0$$

and

$$|e_1\rangle = (q + q^{-1})^{-1}(d^2 - d - 1)^{\frac{1}{2}} \left\{ \left| \underset{\cdot}{\curlyvee} \right| + q \left| \underset{\cdot\cdot}{\sqcup} \right| - q(q+1) \left| \sqcup\sqcup \right| \right\}$$

$$|e_2\rangle = (q + q^{-1})^{-1} d^{-\frac{1}{2}} \left\{ \left| \underset{\cdot}{\curlyvee} \right| - q^{-1} \left| \underset{\cdot\cdot}{\sqcup} \right| - (q-1) \left| \sqcup\sqcup \right| \right\} \tag{96}$$

$$|e_3\rangle = d^{-1} \left| \sqcup\sqcup \right|$$

$$E = \text{Block diag}(0, 0, E_3, 0, 0), \quad z = d^2 - d - 1, \quad d = 1 + q + q^{-1} \quad (q = \text{real}) \tag{97}$$

$$E_A = \begin{bmatrix} 0 & & \\ & 0 & \\ & & d \end{bmatrix}, \quad E_B = \begin{bmatrix} d^{-1}z & -d^{\frac{1}{2}}z^{\frac{1}{2}} & d^{-1}z^{\frac{1}{2}} \\ -d^{\frac{1}{2}}z^{\frac{1}{2}} & 1 & -d^{-\frac{1}{2}} \\ d^{-1}z^{\frac{1}{2}} & -d^{-\frac{1}{2}} & d^{-1} \end{bmatrix} \tag{98}$$

$$A = \begin{bmatrix} q & & \\ & -q^{-1} & \\ & & q^{-2} \end{bmatrix},$$

$$B = \begin{bmatrix} d^{-1}q^4(q+q^{-1})^{-1} & -d^{-\frac{1}{2}}q^{-4}(q+q^{-1})^{-1}Z^{-\frac{1}{2}} & d^{-1}q^{-1}Z^{\frac{1}{2}} \\ -d^{-\frac{1}{2}}q^{-4}(q+q^{-1})^{-1}Z^{-\frac{1}{2}} & (q+q^{-1})^{-1}(d-2) & d^{\frac{1}{2}}q \\ d^{-1}q^{-1}Z^{\frac{1}{2}} & d^{\frac{1}{2}}q & d^{-1}q \end{bmatrix} \tag{99}$$

The 3-D representation of B–W algebra in terms of the topological basis can be Yang–Baxterized to yield solution of YBE.[41]

$$\check{R}(x) = x(x-1)\lambda_1 S^{-1} + x\left(1 + \frac{\lambda_1}{\lambda_2} + \frac{\lambda_1}{\lambda_3} + \frac{\lambda_1^2}{\lambda_2\lambda_3}\right)I - \frac{\lambda_1}{\lambda_2\lambda_3}(x-1)S. \quad (100)$$

When $x = e^{\hbar u}$, $q = e^{-\hbar\frac{\gamma}{2}}$ and replacing u by $u\gamma^{-1}$ the rational limit of $\check{R}(x)$ gives:

$$\check{R}_{\text{rat.}}(u) = I + uP - u\left(u - \frac{S\pm 1}{2}\right)^{-1}F, \quad (\sigma = \pm q^S). \quad (101)$$

The Γ represents the additional contribution to spin-1 model caused by B-W algebra

$$F = I - \frac{1}{2}\frac{\partial(S - S^{-1})}{\partial q}\bigg|_{q=1}. \quad (102)$$

The Hamiltonian for the chain model can then be defined by[41]

$$\hat{H} \sim \frac{\partial\check{R}(u)}{\partial u}\bigg|_{u=0} \sim \sum_i P_{i,i+1} - 2\sum_i F_{i,i+1}, \quad (103)$$

where $P_{i,i+1}$ stands for the permutation. The 9×9 matrix form of F is given by:

$$F = \begin{bmatrix} 0 & 0 & 0 & & \\ & 1\ 1\ 1 & & \\ 0 & 1\ 1\ 1 & 0 & 9\times 9 \\ & 1\ 1\ 1 & & \\ 0 & 0 & 0 & \end{bmatrix} \quad (104)$$

where σ and I represent 3×3 matrices with all the elements 0 or 1, respectively. The result can be easily extended to the S being unitary. Essentially, (104) may describe the quadruple interaction.

9. Conclusion

The topological basis for BWA has been constructed based on the graphic technique in the knot theory that is a little progress to extend the anyon theory associated with T–L algebra.

The velocity additivity of the spectrum parameters for the solution of YBE associated with quantum information obeys Lorentz formula rather than Galileo's ones.

Acknowledgments

The authors thank the collaborators Z. H. Wang, J. L. Chen, S. W. Hu and X. B. Peng. This work is in part supported by NSF of China (Grant No. 10575053).

References

1. M. Jimbo (ed.), *Yang–Baxter Eq. in Integrable Systems* (World Scientific, Singapore, 1990).
2. C. N. Yang and M. L. Ge (eds.), *Braid Group, Knot Theory and Statistical Mechanics* (World Scientific, Singapore, 1990).
3. C. N. Yang, *Phys. Rev. Lett.* **19**, 1312 (1967).
4. C. N. Yang, *Phys. Rev.* **168**, 1920 (1968).
5. R. J. Baxter, *Exactly Solvable Models in Statistic Mechanics* (Academic, New York, 1982).
6. R. J. Baxter, *Ann. Phys. (N.Y.)* **70**, 193 (1972).
7. L. A. Takhtadzhan and L. Faddeev, *Russ. Math. Surv.* **34**, 11 (1979).
8. L. D. Faddeev, *Soviet Sci. Rev. Sect. C: Math. Phys. Rev.* **1**, 107 (1981).
9. P. P. Kulish and E. K. Sklyanin, *Lecture Notes in Physics*, Vol. 151 (Springer, Berlin, 1982), pp. 61–119.
10. P. P. Kulish and E. K. Sklyanin, *J. Soviet Math.* **19**, 1596 (1982).
11. L. Kauffman and S. Lomonaco Jr., *New J. Phys.* **6**, 134 (2004).
12. J. Preskill, *Lecture Notes for Physics 219: Quantum Computation* (2004). http://www.theory.caltech.edu/preskill/ph229/.
13. Z. Wang, Topologization of electron liquids with Chern–Simons theory and quantum computation, *Differential Geometry and Physics*, Nankai Tracts. Math. Vol. 10 (World Scientific, Hackensack, 2006), pp. 106–120.
14. M. H. Freedman, M. Larsen and Z. Wang, *Commun. Math. Phys.* **227**, 605 (2002).
15. S. Das Sarma, M. Freedman and C. Nayak, *Phys. Rev. Lett.* **94**, 166802 (2005).
16. Z. Wang, *Nankai Lectures on TQFT* (June 5–7, 2006).
17. C. Nayak and F. Wilczek, *Nucl. Phys. B* **479**, 529 (1996).
18. C. Nayak *et al.*, *Rev. Mod. Phys.* **80**, 1083 (2008).
19. J. K. Slingerland and F. A. Bais, *Nucl. Phys. B* **612[FS]**, 229 (2001).
20. E. H. Rezayi and N. Read, *Nucl. Phys. B* **56**, 16864 (1996).
21. Y. Zhang, L. Kauffman and M. Ge, *Quantum Inf. Process.* **4**, 159 (2005).
22. J. L. Chen, K. Xue and M. L. Ge, *Phys. Rev. A* **76**, 042324 (2007).
23. S. W. Hu, K. Xue and M. L. Ge, *Phys. Rev. A* **78**, 022319 (2008).
24. J. L. Chen, K. Xue and M. L. Ge, *Ann. Phys.* **323**, 2614 (2008).
25. M. G. Hu, K. Xue and M. L. Ge, *Phys. Rev. A* **78**, 052324 (2008).
26. K. Niu *et al.*, *J. Phys. A* **44**, (2011).
27. L. Takhtadzhan and L. Faddeev, *Russ. Math. Surv.* **34**, 11 (1979).
28. V. F. R. Jones, *Int. J. Mod. Phys. A* **6**, 2035 (1991).
29. M. L. Ge, Y. S. Wu and K. Xue, *Int. J. Mod. Phys. A* **6**, 3735 (1991).
30. H. N. V. Temperley and E. H. Lieb, *Proc. R. Soc. Lond. Ser. A* **322**, 251 (1971).
31. J. Birman and H. Wenzl, *Trans. A. M. S.* **313**, 249 (1989).
32. J. Murakami, *Osaka J. Math.* **24**, 745 (1987).
33. H. Wenzl, *Ann. Math.* **128**, 179 (1988).
34. L. Kauffman, *Knots in Physics* (World Scientific, Singapore, 1991).
35. A. Benvegnu and M. Spera, *Rev. Math. Phys.* **18**, 10 (2006).
36. D. Donoho, *IEEE Trans. Inform. Theory* **52**, 1289 (2006).
37. E. Cande's and T. Tao, *IEEE Trans. Inform. Theory* **52**, 5406 (2006).
38. R. Boraniuk, J. Romberg and M. Wakin, http://www.dsp.ece.rice.edu/richb/~alks/cs-tutorial-ITA-feb08-complet.pdf
39. M. Jimbo, *Commun. Math. Phys.* **102**, 537 (1986).
40. M. L. Ge and A. C. T. Wu, *J. Phys. A* **25**, L807 (1992).
41. Y. Cheng, M. L. Ge and K. Xue, *Commun. Math. Phys.* **136**, 195 (1991).

Chapter 8

NONDISTILLABLE ENTANGLEMENT GUARANTEES DISTILLABLE ENTANGLEMENT

LIN CHEN

Department of Pure Mathematics and Institute for Quantum Computing,
University of Waterloo, Waterloo, Ontario, N2L 3G1, Canada
Centre for Quantum Technologies, National University of Singapore,
3 Science Drive 2, Singapore 117542
cqtcl@nus.edu.sg

MASAHITO HAYASHI

Graduate School of Mathematics, Nagoya Unversity,
Furocho, Chikusaku, Nagoya 464-860, Japan
Centre for Quantum Technologies, National University of Singapore,
3 Science Drive 2, Singapore 117542
masahito@math.nagoya-u.ac.jp

The monogamy of entanglement is one of the basic quantum mechanical features, which says that when two partners Alice and Bob are more entangled then either of them has to be less entangled with the third party. Here we qualitatively present the converse monogamy of entanglement: given a tripartite pure system and when Alice and Bob are entangled and nondistillable, then either of them is distillable with the third party. Our result leads to the classification of tripartite pure states based on bipartite reduced density operators, which is a novel and effective way to this long-standing problem compared to the means by stochastic local operations and classical communications. Furthermore we systematically indicate the structure of the classified states and generate them. We also extend our results to multipartite states.

Keywords: Multipartite state; entanglement; separability, partial transpose; reduction criterion; majorization; conditional entropy.

1. Introduction

The monogamy of entanglement is a purely quantum phenomenon in physics[1] and has been used in various applications, such as bell inequalities[2] and quantum security.[3] In general, it indicates that the more entangled the composite system of two partners Alice (A) and Bob (B) is, the less entanglement between A (B) and the environment E there is. The security of many quantum secret protocols can be guaranteed quantitatively.[3,4] However the converse statement generally does not hold, namely when A and B are less entangled, we cannot decide whether A (B) and E are more entangled. In fact even when the formers are classically correlated namely separable,[5] the latters may be also separable. For example, this is realizable by the tripartite Greenberger-Horne-Zeilinger (GHZ) state.

Nevertheless, it is still important to *qualitatively* characterize the above converse statement in the light of the hierarchy of entanglement of bipartite systems. Such a characterization defines a converse monogamy of entanglement, and there is no classical counterpart. Besides, it is also expected to be helpful for treating a quantum multi party protocol when the third party helps the remaining two parties, for it guarantees the property of one reduced density operator from another. To justify the hierarchy of entanglement, we recall six well-known conditions, i.e., the separability, positive-partial-transpose (PPT),[6,7] nondistillability of entanglement under local operations and classical communications (LOCC),[7] reduction property (states satisfying reduction criterion),[8] majorization property[9] and negativity of conditional entropy.[8] These conditions form a hierarchy since a bipartite state satisfying the former condition will satisfy the latter too. Therefore, the strength of entanglement in the states satisfying the conditions in turn becomes gradually *weak*.

For example, the hierarchy is closely related to the distillability of entanglement.[7] While PPT entangled states cannot be distilled to Bell states for implementing quantum information tasks, Horodecki's protocol can distill a state that violates reduction criterion.[8] That is, the former entangled state is useless as a resource while the latter entangled state is useful. So the usefulness of entangled states can be characterized by this hierarchy. Recently, a hierarchy of entanglement has been developed based on these criteria.[10]

In this paper, for simplicity we consider four most important conditions, namely the separability, PPT, nondistillability and the reduction criteria. Then we establish a hierarchy of entanglement consisting of five sets: separable states (S), nonseparable PPT states (P), nonPPT nondistillable states (N), distillable reduction states

Fig. 1.   Hierarchy of bipartite states in terms of five sets S, P, N, D, M. Intuitively, the sets S and P form all PPT states, the sets S, P, N, D and M form all states satisfying and violating the reduction criterion, respectively. So the five sets constitute the set of bipartite states and there is no intersection between any two sets. The strength of entanglement of the five sets becomes weak in turn, $S \leq P \leq N \leq D \leq M$.

(D), and nonreduction states (M), see Fig. 1. In particular the states belonging to M are always distillable.[8]

We show that when the entangled state between A and B, i.e., ρ_{AB} belongs to the set D, then the state between A (B) and E, i.e., $\rho_{AE}(\rho_{BE})$ belongs to the set D or M. Likewise when ρ_{AB} belongs to P or N, then $\rho_{AE}(\rho_{BE})$ must belong to M. Hence we can qualitatively characterize the converse monogamy of entanglement as follows: when the state ρ_{AB} is weakly entangled, then ρ_{AE} is generally strongly entangled in terms of the five sets S, P, N, D, M. These assertions follow from a corollary of Theorem 2.1 to be proved later.

Theorem 1.1. *Suppose a tripartite pure state has a nondistillable bipartite reduced state. Then another bipartite reduced state is separable if and only if it satisfies the reduction criterion.*

From this theorem, we will solve two conjectures on the existence of specified tripartite state proposed by Thapliyal in 1999.[11] On the other hand, the theorem also helps develop the classification of tripartite pure states based on the three reduced states, each of which could be in one of the five cases S, P, N, D, M. So there are at most $5^3 = 125$ different kinds of tripartite states. Evidently, some of them do not exist due to Theorem 1.1. It manifests that the quantum behavior of a global system is strongly restricted by local systems. By generalizing to many-body systems, we can realize the quantum nature on macroscopic size in terms of the microscopic physical systems. This is helpful to the development of matter and material physics.[12] Hence, in theory it becomes important to totally identify different tripartite states.

To explore the problem, we describe the properties for reduced states ρ_{AB}, ρ_{BC}, and ρ_{CA} of the state $|\Psi\rangle$ by X_{AB}, X_{BC}, X_{CA} that take values in S, P, N, D, M. The subset of such states $|\Psi\rangle$ is denoted by $S_{X_{AB}X_{BC}X_{CA}}$, and the subset is nonempty when there exists a tripartite state in it. For example, the GHZ state belongs to the subset S_{SSS}. Furthermore as is later shown in Table 1, $|\Psi\rangle$ belongs to the subset S_{SSM} when the reduced state ρ_{CA} is an entangled maximally correlated state.[13] By Theorem 1.1, one can readily see that any nonempty subset is limited in nine essential subsets,

$$S_{SSS}, S_{SSM}, S_{SMM}, S_{PMM}, S_{NMM}$$

and

$$S_{DDD}, S_{DDM}, S_{DMM}, S_{MMM}.$$

Hence up to permutation, the number of nonempty subsets for tripartite pure states is at most $21 = 1 \times 3 + 3 \times 6$. In particular, it is a long-standing open problem that whether S_{NMM} exists.[17] Except the subsets generated from S_{NMM}, we will demonstrate that the rest 18 subsets are indeed nonempty by explicit examples. These subsets are not preserved under the conventional classification by the invertible stochastic LOCC (SLOCC).[14-16] We will explain these results in Sec. 3.

Table 1.   Classification of tripartite states $|\psi\rangle_{ABC}$ in terms of the bipartite reduced states. The table contains neither the classes generated from the permutation of parties, and nor the subset $\mathcal{S}_{MMM}$ since for which there is no fixed relation between the tensor rank rk(ψ) and *local ranks* $d_A(\Psi)$, $d_B(\Psi)$, and $d_C(\Psi)$. They are simplified to r, d_A, d_B, and d_C when there is no confusion. All expressions are up to local unitaries (LU) and all sums run from 1 to r. In the subset $\mathcal{S}_{SMM}$, the linearly independent states $|c_i\rangle$ are the support of space $\mathcal{H}_C$. Apart from the subset $\mathcal{S}_{NMM}$, all other eight subsets are nonempty in terms of specific examples in Sec. 3.

$\mathcal{S}_{X_{AB}X_{BC}X_{CA}}$	$\mathcal{S}_{SSS}$	$\mathcal{S}_{SSM}$	$\mathcal{S}_{SMM}$
expression of $|\psi\rangle_{ABC}$	$\sum_i \sqrt{p_i}|iii\rangle$	$\sum_i \sqrt{p_i}|i,b_i,i\rangle$	$\sum_i \sqrt{p_i}|a_i,b_i,c_i\rangle$
tensor rank and local ranks	$r = d_A = d_B = d_C$	$r = d_A = d_C \geq d_B$	$r = d_C \geq d_A, d_B$

$\mathcal{S}_{X_{AB}X_{BC}X_{CA}}$	$\mathcal{S}_{PMM}$	$\mathcal{S}_{NMM}$	$\mathcal{S}_{DDD}$
expression of $|\psi\rangle_{ABC}$	$\sum_i \sqrt{p_i}|a_i,b_i,c_i\rangle$	$\sum_i \sqrt{p_i}|a_i,b_i,c_i\rangle$	$\sum_i \sqrt{p_i}|a_i,b_i,c_i\rangle$
tensor rank and local ranks	$r \geq d_C > d_A, d_B$	$r \geq d_C > d_A, d_B$	$r > d_C = d_B = d_A$

$\mathcal{S}_{X_{AB}X_{BC}X_{CA}}$	$\mathcal{S}_{DDM}$	$\mathcal{S}_{DMM}$
expression of $|\psi\rangle_{ABC}$	$\sum_i \sqrt{p_i}|a_i,b_i,c_i\rangle$	$\sum_i \sqrt{p_i}|a_i,b_i,c_i\rangle$
tensor rank and local ranks	$r > d_C = d_A \geq d_B$	$r \geq d_C \geq d_A, d_B$ $r > d_A, d_B$

Furthermore, we show that the subsets form a commutative monoid and it is a basic algebraic concept. We systematically characterize the relation of the subsets by generating them under the rule of monoid in the late part of Sec. 3.

We also extend our results to multipartite scenario. In particular, we introduce the multipartite separable, PPT and nondistillable states. They become pairwise equivalent when they are compatible to a pure state, see Theorem 4.1, Sec. 4. Finally we conclude in Sec. 5.

2. Unification of Entanglement Criterion

In quantum information, the following six criteria are extensively useful for studying bipartite states ρ_{AB} in the space $\mathcal{H}_A \otimes \mathcal{H}_B$.

(1) Separability: ρ_{AB} is the convex sum of product states.[5]
(2) PPT condition: the partial transpose of ρ_{AB} is semidefinite positive.[6]
(3) Nondistillability: no pure entanglement can be asymptotically extracted from ρ_{AB} under LOCC, no matter how many copies are available.[7]
(4) Reduction criterion: $\rho_A \otimes I_B \geq \rho_{AB}$ and $I_A \otimes \rho_B \geq \rho_{AB}$.[8]
(5) Majorization criterion: $\rho_A \succ \rho_{AB}$ and $\rho_B \succ \rho_{AB}$.[9]
(6) Conditional entropy criterion: $H_\rho(B|A) = H(\rho_{AB}) - H(\rho_A) \geq 0$ and $H_\rho(A|B) = H(\rho_{AB}) - H(\rho_B) \geq 0$, where H is the von Neumann entropy.

It is well-known that the relation $(1) \Rightarrow (2) \Rightarrow (3) \Rightarrow (4) \Rightarrow (5) \Rightarrow (6)$ holds for any state ρ_{AB}.[7-9] In particular apart from $(2) \Rightarrow (3)$ whose converse is a famous open problem,[17] all other relations are strict. We will show that these conditions become equal when we further require ρ_{BC} is nondistillable. First, under this restriction the conditions (5) and (6) are respectively simplified into $(5')$ $\rho_A \sim \rho_{AB}$, where $\sim$ denotes that ρ_A and ρ_{AB} have identical eigenvalues, and $(6')$ $H(\rho_A) = H(\rho_{AB})$. Second, when ρ_{BC} is nondistillable, since $\rho_{AB} \succ \rho_A$ holds, the above two conditions $(5')$ and $(6')$ are equivalent. Now we have:

Theorem 2.1. *For a tripartite state* $|\Psi\rangle_{ABC}$ *with a nondistillable reduced state* ρ_{BC} *namely condition (3), then conditions (1)–(6), $(5')$ and $(6')$ are equivalent for* ρ_{AB}.

The proof is given in the appendix of this paper. We can readily get Theorem 1.1 from Theorem 2.1, and provide its operational meaning as the main result of the work.

Theorem 2.2. (Converse monogamy of entanglement). *Consider a tripartite state* $|\Psi\rangle_{ABE}$ *with entangled reduced states* ρ_{AB}, ρ_{AE} *and* ρ_{BE}. *When* ρ_{AB} *is a weakly entangled state P or N (D), the states ρ_{AE} and ρ_{BE} are strongly entangled states M $(D$ or $M)$.*

To our knowledge, the converse monogamy of entanglement is another basic feature of quantum mechanics and there is no classical counterpart since classical correlation can only be "quantified". In contrast, quantum entanglement has qualitatively different levels of strength and they have essentially different usefulness from each other. For example the states in the subset N cannot be distilled while those in M are known to be distillable.[8] So only the latter can directly serve as an available resource for quantum information processing and it implies the following paradox. **A useless entangled state between A and B strengthens the usefulness of entanglement resource between A (B) and the environment.** Therefore, the converse monogamy of entanglement indicates a dual property to the monogamy of entanglement. Not only the amount of entanglement, the usefulness of entanglement in a composite system is also restricted by each other.

Apart from bringing about the converse monogamy of entanglement, Theorem 2.1 also promotes the study over a few important problems. For instance, the equivalence of (1) and (2) is a necessary and sufficient condition of deciding separable states, beyond that for states of rank not exceeding four.[6,18,19] Besides, the equivalence of (2) and (3) indicates another kind of nonPPT entanglement activation by PPT entanglement.[20] For later convenience, we explicitly work out the expressions of states satisfying the assumptions in Theorem 2.1.

Lemma 2.1. *The tripartite pure state with two nondistillable reduced states* ρ_{AB} *and* ρ_{AC}, *if and only if it has the form* $\sum_i \sqrt{p_i}|b_i, i, i\rangle$ *upto local unitary operators. In other words, the reduced state* ρ_{BC} *is maximally correlated.*

For the proof see Lemma 11 in Ref. 19. We apply our results to handle two open problems in Fig. 4 of Ref. 11, i.e., whether there exist tripartite states with two PPT bound entangled reduced states, and tripartite states with two separable and one bound entangled reduced states. Here we give negative answers to these open problems in terms of Theorem 2.1 and Lemma 2.1. As the first problem is trivial, we account for the second conjecture. Because the required states have the form $\sum_{i=1}^{d} \sqrt{p_i}|ii\rangle|c_i\rangle$, where ρ_{AC} and ρ_{BC} are separable. So the reduced state ρ_{AB} is a maximally correlated state, which is either separable or distillable. It readily denies the second problem.

It is noticeable that the converse of Theorem 2.1 does not hold. That is, for a tripartite state $|\Psi\rangle_{ABC}$ for which conditions (1)–(6), (5') and (6') are equivalent for ρ_{AB}, the reduced state ρ_{BC} is not necessarily nondistillable. An example is the state $|000\rangle + (|0\rangle + |1\rangle)|11\rangle$.

Finally we extend Theorem 2.1 for the tripartite state containing a qubit reduced state. For this purpose we introduce the following known result [Ref. 21, Remark 1].

Lemma 2.2. *A $2 \times N$ state is PPT if and only if it satisfies the reduction criterion.*

Based on it we have:

Theorem 2.3. *Suppose $|\Psi\rangle_{ABC}$ contains a qubit reduced state. If ρ_{BC} satisfies condition (4), then conditions (1)–(6), (5') and (6') are equivalent for ρ_{AB}.*

Proof. When ρ_{BC} contains a qubit reduced state, the claim follows from Lemma 2.2 and Theorem 2.1. So it suffices to consider the case rank $\rho_A = 2$. Since ρ_{BC} satisfies condition (4), we obtain rank $\rho_A \geq$ rank ρ_B, rank ρ_C. In other word ρ_{BC} is a two-qubit state and the claim again follows from Lemma 2.2 and Theorem 2.1. This completes the proof.   □

For tripartite states with higher dimensions, Theorem 2.3 does not apply anymore and we will see available examples in next section. As the concluding remark of this section, we propose the following conjecture.

Conjecture 2.1. *For a tripartite $3 \times 3 \times 3$ state $|\Psi\rangle_{ABC}$, suppose ρ_{BC} satisfies (4) and ρ_{AB} satisfies (5). Then ρ_{AB} also satisfies (4).*

3. Classification with Reduced States

Theorem 2.2 says that the quantum correlation between two parties of a tripartite system is dependent on the third party. From Theorem 2.1 and the discussion to conjectures in Ref. 11, we can see that the tripartite pure state with some specified bipartite reduced states may not exist. This statement leads to a classification of tripartite states in terms of the three reduced states.[22] As a result, we obtain the different subsets of tripartite states in Table 1 in terms of tensor rank and local ranks of each one-party reduced state. In the language of quantum information, the tensor rank of a multipartite state, also known as the Schmidt measure of entanglement,[23] is equal to the least number of product states to expand this state. For instance, any multiqubit GHZ state has tensor rank two. So tensor rank is bigger or equal to any local rank of a multipartite pure state. As tensor rank is invariant under invertible SLOCC,[15] it has been widely applied to classify SLOCC-equivalent multipartite states recently.[16]

Here we will see that, tensor rank is also essential to the classification in Table 1. We will justify the statement for each subset in Table 1, then we show their nonemptyness by proposing specific examples.

First the statement for the subsets S_{SSS}, S_{SSM} and S_{SMM} follows from Lemma 2.1, and Lemma 2 in Ref. 24, respectively. The nonemptyness readily follows from the states $|\psi\rangle_{ABC}$ in Table 1. To verify the statement for S_{PMM} and S_{NMM}, we propose the following result

Lemma 3.1. *A $M \times N$ state with rank N is nondistillable if and only if it is separable and is the convex sum of just N product states, i.e., $\sum_{i=1}^{N} p_i |a_i, b_i\rangle\langle a_i, b_i|$.*

Proof. It suffices to show the necessity. Let ρ be a $M \times N$ state with rank N and suppose it is nondistillable. From Lemma 11 in Ref. 19 we obtain ρ is PPT. The claim then follows from Ref. 18. This completes the proof. □

As a result, the PPT entangled state ρ_{AB} satisfies $d_A, d_B < \operatorname{rank}\rho_{AB}$. So the purification of PPT entangled states ρ_{AB} is subject to the statement for subset S_{PMM} in Table 1. It also shows the nonemptyness of S_{PMM}. A similar argument can be used to justify the statement for S_{NMM} in Table 1, if it really exists.

Next we study the subset S_{DDD}. Since the reduced state ρ_{AB} satisfies the reduction criterion, we have $d_C \geq d_A, d_B$. By applying the same argument to other reduced states we obtain $d_A = d_B = d_C$. Since ρ_{AB} is distillable, the rest statement $r > d_A$ in Table 1 follows from the following observation.

Lemma 3.2. *Assume that $\operatorname{rk}(\Psi) = \max\{d_A, d_B\}$. Then the conditions (1)–(4) are equivalent for ρ_{AB}.*

Proof. It suffices to show that the state ρ_{AB} is separable when it satisfies the reduction criterion. Let $|\Psi\rangle = \sum_{i=1}^{\operatorname{rk}(\Psi)} \sqrt{p_i}|a_i, b_i, c_i\rangle$, and V_A an invertible matrix such that $V_A|i\rangle = |a_i\rangle$. Now, we focus on the pure state $|\Psi'\rangle = K V_A^{-1} \otimes I_B \otimes I_C |\Psi\rangle$, where K is the normalized constant. Then the reduced state ρ'_{AB} satisfies $\rho'_A \otimes I_B \geq \rho'_{AB}$, and hence $\rho'_A \succ \rho'_{AB}$.[9] Since ρ'_{BC} is separable, we have $\rho'_A \sim \rho'_{AB}$. So the state ρ'_{AB}, and equally ρ_{AB} is separable in terms of Theorem 2.1. □

Example 3.1. It is noticeable that under the same assumption in Lemma 3.2, the equivalence between conditions (1)–(5) does not hold. As a counterexample, we consider the symmetric state $|\Psi\rangle = 1/(\sqrt{r+7})(\sum_{i=2}^{r} |iii\rangle + (|1\rangle + |2\rangle)(|1\rangle +$

$|2\rangle)(|1\rangle + |2\rangle))$. It is symmetric and thus satisfies condition (5). On the other hand one can directly show that any reduced state of $|\Psi\rangle$ violate the reduction criterion, so it does not satisfy conditions (1)–(4). Hence $|\Psi\rangle$ belongs to the subset $\mathcal{S}_{MMM}$. It indicates that tensor rank alone is not enough to characterize the hierarchy of bipartite entanglement.

Example 3.2. The symmetric state $|\psi_r\rangle = 1/(\sqrt{2r})(|312\rangle + |123\rangle + |231\rangle + |213\rangle + |132\rangle + |321\rangle) + (1/\sqrt{r})\sum_{j=4}^{r}|jjj\rangle$ indicates the nonemptyness of $\mathcal{S}_{DDD}$ for $d_A \geq 3$. To see it, it suffices to show one of the reduced states, say ρ_{AB} satisfies the reduction criterion and is distillable simultaneously. The former can be directly justified. By performing the local projector $P = |1\rangle\langle 1| + |2\rangle\langle 2|$ on system A, we obtain the resulting state $P \otimes I\rho_{AB}P \otimes I = (|12\rangle + |21\rangle)(\langle 12| + \langle 21|)$ which is a Bell state. So ρ_{AB} is distillable.

However there is no state with $d_A = 2$ in $\mathcal{S}_{DDD}$. The reason is that a two-qubit state satisfying the reduction criterion is also PPT by Lemma 2.2. Hence it must be separable in terms of Peres' condition.[25] It contradicts with the statement that ρ_{AB} is distillable.

By using a similar argument to $\mathcal{S}_{DDD}$, one can verify the statement for $\mathcal{S}_{DDM}$ in Table 1. A concrete example will be built by the rule of monoid later. Hence $\mathcal{S}_{DDM}$ is nonempty.

Third, we characterize the subset $\mathcal{S}_{DMM}$ by the tensor rank of states in this subset. It follows from the definition of reduction criterion that $d_C \geq d_A, d_B$. This observation and Lemma 3.2 justify the statement for $\mathcal{S}_{DMM}$ in Table 1. In order to show the tightness of these inequalities, we consider the nonemptyness of the subset $\mathcal{S}_{DMM}$ with the boundary types $r = d_C > d_A = d_B$ and $r > d_C = d_A = d_B$.

Example 3.3. To justify the former type, it suffices to consider the state $|\psi_a\rangle = (1/\sqrt{6 + 3a^2})(|123\rangle + |231\rangle + |312\rangle + |21\rangle(|3\rangle + a|6\rangle) + |13\rangle(|2\rangle + a|5\rangle) + |32\rangle(|1\rangle + a|4\rangle))$, $a \in R$. It obviously fulfils $r = d_C > d_A = d_B$, so the reduced states ρ_{AC} and ρ_{BC} violate the reduction criterion. Next we focus on the reduced state ρ_{AB}. By performing the local projector $P = |1\rangle\langle 1| + |2\rangle\langle 2|$ on system A, we obtain the resulting state $P \otimes I\rho_{AB}P \otimes I = (|12\rangle + |21\rangle)(\langle 12| + \langle 21|) + a^2|21\rangle\langle 21|$. This is an entangled two-qubit state and is hence distillable.[26] On the other hand to see that ρ_{AB} fulfils the reduction criterion for any a, one need to notice the facts $\rho_A = \rho_B = (1/3)I$ and the eigenvalues of ρ_{AB} are not bigger than $1/3$.

In addition, an example of the latter type is constructed by the rule of monoid later. Thus, we can confirm the tightness of the above inequalities of two boundary types for $\mathcal{S}_{DMM}$.

To conclude, we have verified the statement and existence of all essential subsets in Table 1 except the $\mathcal{S}_{NMM}$.

Comparison to SLOCC classification. We know that there are much efforts towards the classification of multipartite state by invertible SLOCC.[14–16] Hence, it is necessary to clarify the relation between this method and the classification by reduced states in Table 1. When we adopt the former way we have no clear characterization to the hierarchy of bipartite entanglement between the involved parties, i.e., the structure of reduced states becomes messy under SLOCC. Our classification resolves this drawback. Another potential advantage of our idea is that we can apply the known fruitful results of bipartite entanglement, such as the hierarchy of entanglement to further study the classification problem.

Here we explicitly exemplify that the invertible SLOCC only partially preserves the classification in Table 1. We focus on the orbit $\mathcal{O}_{r=d_A=d_B=d_C} := \{|\Psi\rangle | \, \mathrm{rk}(\Psi) = d_A(\Psi) = d_B(\Psi) = d_C(\Psi)\}$, which has nonempty intersection with the subsets $\mathcal{S}_{SSS}$, $\mathcal{S}_{SSM}$ and $\mathcal{S}_{SMM}$. Further, since the subset $\mathcal{S}_{MMM}$ contains the state $|\Psi_a\rangle$, it also has nonempty intersection with $\mathcal{O}_{r=d_A=d_B=d_C}$. Since all state in $\mathcal{O}_{r=d_A=d_B=d_C}$ can be converted to GHZ state by invertible SLOCC, it does not preserve the classification by reduced states. However, the subsets $\mathcal{S}_{DDM}$ and $\mathcal{S}_{DDD}$ are not mixed with $\mathcal{S}_{SSS}$, $\mathcal{S}_{SSM}$ and $\mathcal{S}_{SMM}$ by invertible SLOCC.

Monoid structure. To get a further understanding of Table 1 from the algebraic viewpoint, we define the direct sum for subsets by $\mathcal{S}_{X_{AB}X_{BC}X_{CA}} \mathcal{S}_{Y_{AB}Y_{BC}Y_{CA}} := \mathcal{S}_{\max\{X_{AB},Y_{AB}\} \max\{X_{BC},Y_{BC}\} \max\{X_{CA},Y_{CA}\}}$, where $\max\{X,Y\}$ is the larger one between X and Y in the order $S \leq P \leq N \leq D \leq M$. Therefore when $|\Psi_1\rangle \in \mathcal{S}_{X_{AB}X_{BC}X_{CA}}$ and $|\Psi_2\rangle \in \mathcal{S}_{Y_{AB}Y_{BC}Y_{CA}}$, the state $|\Psi_1 \cdot \Psi_2\rangle := |\Psi_1\rangle \oplus |\Psi_2\rangle \in \mathcal{S}_{\max\{X_{AB},Y_{AB}\} \max\{X_{BC},Y_{BC}\} \max\{X_{CA},Y_{CA}\}}$. This product is commutative and in the direct sum, the subset $\mathcal{S}_{SSS}$ is the unit element but no inverse element exists. So the family of nonempty sets $\mathcal{S}_{X_{AB}X_{BC}X_{CA}}$ with the direct sum is an abelian monoid, which is a commutative semigroup associated with the unit.

The above analysis provides a systematic method to produce the subsets in the monoid structure, except $\mathcal{S}_{NMM}$ whose existence is an open problem so far. Generally we have $\mathcal{S}_{SMM} = \mathcal{S}_{SSM}\mathcal{S}_{SMS}, \mathcal{S}_{DDM} = \mathcal{S}_{DDD}\mathcal{S}_{SSM}, \mathcal{S}_{DMM} = \mathcal{S}_{DDD}\mathcal{S}_{SMM}$ and $\mathcal{S}_{MMM} = \mathcal{S}_{PMM}\mathcal{S}_{MSS}$. So it is sufficient to check the nonemptyness of subsets $\mathcal{S}_{SSS}, \mathcal{S}_{SSM}, \mathcal{S}_{PMM}, \mathcal{S}_{DDD}$. This fact has been verified in Sec. III, and we can use the method to produce states in $\mathcal{S}_{DMM}$. The following is an example.

If we choose $|\Psi_1\rangle \in \mathcal{S}_{SMM}$ and $|\Psi_2\rangle \in \mathcal{S}_{DDD}$ and both have $d_A = d_B = d_C$. Then the state $|\Psi_1 \cdot \Psi_2\rangle$ verifies the nonemptiness of the subset $\mathcal{S}_{DMM}$ with the condition $r > d_C = d_A = d_B$. The arguments have verified the existence of a boundary type mentioned in the second paragraph below Lemma 3.2.

4. Generalization to Multipartite System

We begin by introducing the following results from Refs. 11, 27. By fully separable states ρ of N-partite systems, we mean $\rho = \sum_i p_i \rho_{1,i} \otimes \cdots \otimes \rho_{N,i}$.

Lemma 4.1. *The $M \times N$ states of rank less than M or N are distillable, and consequently they are NPT.*

Lemma 4.2. *The N-partite state $|\psi\rangle$ has N fully separable $(N-1)$-partite reduced states if and only if $|\psi\rangle$ is a generalized GHZ state $\sum_i \sqrt{p_i}|ii\cdots i\rangle$ up to LU.*

Next, we generalize Lemma 2.1 and 4.2. For this purpose we define the *n-partite nondistillable state* $\rho_{1\cdots n}$, in the sense that one cannot distill any pure entangled state by collective LOCC over any bipartition of parties $1,\ldots,m : (m+1),\ldots,n$. By "collective LOCC" we regard parties $1,\ldots,m$ and $(m+1),\ldots,n$ as two local parties, respectively. Similarly, we define the *n-partite PPT state* in the sense that any bipartition of the state is PPT. Hence, such states contain a more restrictive quantum correlation than the general multipartite state. Evidently, the multipartite PPT state is a special multipartite nondistillable state. The converse is unknown even for the bipartite case.[17] In what follows we will show the equivalence of multipartite nondistillable, PPT and fully separable states which are reduced states of a multipartite pure state.

For convenience we denote $\rho_{\bar{i}}$ as a $(N-1)$-partite state by tracing out the party A_i from the N-partite state $|\psi\rangle$, i.e., $\rho_{\bar{i}} = \mathrm{Tr}_i |\psi\rangle\langle\psi|$. With these definitions we have:

Theorem 4.1. *The following four statements are equivalent for a N-partite state $|\psi\rangle$.*

(1) *$|\psi\rangle$ has n nondistillable $(N-1)$-partite reduced states $\rho_{\bar{1}},\ldots,\rho_{\bar{n}}$, $N \geq n \geq 2$;*
(2) *$|\psi\rangle$ has n PPT $(N-1)$-partite reduced states $\rho_{\bar{1}},\ldots,\rho_{\bar{n}}$, $N \geq n \geq 2$;*
(3) *$|\psi\rangle$ has n fully separable $(N-1)$-partite reduced states $\rho_{\bar{1}},\ldots,\rho_{\bar{n}}$, $N \geq n \geq 2$;*
(4) *$|\psi\rangle = \sum_i \sqrt{p_i}|i\rangle^{\otimes n}|b_{i,n+1}\rangle \otimes \cdots \otimes |b_{i,N}\rangle$ up to LU.*

Proof. Suppose $|\psi\rangle$ is of dimensions $d_1 \times \cdots \times d_N$. The direction (4) $\rightarrow$ (3) $\rightarrow$ (2) $\rightarrow$ (1) is evident. To show (1) $\rightarrow$ (4), suppose $\rho_{\bar{1}},\ldots,\rho_{\bar{n}}$ are nondistillable. By using Lemma 4.1 and the definition of multipartite nondistillable states, we can obtain $d := d_1 = \cdots = d_n \geq d_{n+1},\ldots,d_N$. By combining the parties $A_3,\ldots,A_N$ into one party, we obtain a tripartite state satisfying Lemma 2.1. Hence we have $|\psi\rangle = \sum_{i=1}^d \sqrt{p_i}|ii\rangle|\varphi_i\rangle$ where $|\varphi_i\rangle \in \bigotimes_{i=3}^N \mathcal{H}_i$.

We show that $|\varphi_i\rangle$ are fully product states. The proof is by contradiction. Suppose there is, say $|\varphi_1\rangle$ which is not fully factorized. So we can write it as a bipartite entangled state in the space $\mathcal{H}_C \otimes \mathcal{H}_D$. In other word we have the 4-partite state $|\psi\rangle \in \mathcal{H}_1 \otimes \mathcal{H}_2 \otimes \mathcal{H}_C \otimes \mathcal{H}_D$ and the tripartite reduced state $\sigma_{2,C,D} = \sum_{i=1}^d |i\rangle\langle i|_2 \otimes |\varphi_i\rangle\langle\varphi_i|_{CD}$. By performing the projector $|1\rangle\langle 1|$ on space $\mathcal{H}_2$, we can distill a pure entangled state $|\varphi_1\rangle$ from $\sigma_{2,C,D}$. It contradicts with the assumption on $\rho_{\bar{1}},\ldots,\rho_{\bar{n}}$. So every state $|\varphi_i\rangle$ has to be fully factorized and up to LU, we have

$$|\psi\rangle = \sum_{i=1}^d \sqrt{p_i}|ii\rangle \bigotimes_{j=3}^N |a_{i,j}\rangle. \tag{4.1}$$

Next, we combine parties $A_1, A_4, \ldots, A_N$ together and make $|\psi\rangle$ a new tripartite state. Because $\rho_{\bar{3}}$ is nondistillable and any entangled maximally correlated state is distillable,[13] the states $|a_{i,3}\rangle$ have to be orthonormal. In a similar way we can prove the states $|a_{i,j}\rangle$, $j = 4, \ldots, n$ are orthonormal, respectively. So we have justified the statement. This completes the proof of (1) $\to$ (4). So all four statements (1), (2), (3), (4) are equivalent. □

The following result is a stronger version of Lemma 4.2.

Corollary 4.1. *The N-partite state $|\psi\rangle$ has N nondistillable $(N-1)$-partite reduced states if and only if it is a generalized GHZ state $|\psi\rangle = \sum_i \sqrt{p_i}|i\rangle^{\otimes N}$ up to LU.*

5. Conclusions

We have proposed the converse monogamy of entanglement such that when Alice and Bob are weakly entangled, then either of them is generally strongly entangled with the third party. We believe that the converse monogamy of entanglement is an essential quantum mechanical feature and it promises a wide application in deciding separability, entanglement distillation and quantum cryptography. Our result presents two main open questions: First, can we propose a concrete quantum scheme by the converse monogamy of entanglement? Such a scheme will indicate a new essential difference between the classical and quantum rules, just like that from quantum cloning[28] and the negative conditional entropy.[29] Second, different from the monogamy of entanglement which relies on the specific entanglement measures,[3] the converse monogamy of entanglement only relies on the strength of entanglement. So can we get a better understanding by adding other criteria on the strength of entanglement such as the nondistillability?

We also have shown tripartite pure states can be sorted into 21 subsets and they form an abelian monoid. It exhibits a more canonical and clear algebraic structure of tripartite system compared to the conventional SLOCC classification.[14] More efforts from both physics and mathematics are required to understand such structure.

6. Acknowledgments and Appendices

We thank Fernando Brandao for helpful discussion on the reduction criterion and Andreas Winter for reading through the paper. The CQT is funded by the Singapore MoE and the NRF as part of the Research Centres of Excellence programme. MH is partially supported by a MEXT Grant-in-Aid for Young Scientists (A) No. 20686026 and a MEXT Grant-in-Aid for Scientific Research (A) No. 23246071.

Appendix

We prove Theorem 2.1 based on the following preliminary lemma.

Lemma 6.1. *Consider a tripartite state $|\Psi_{ABC}\rangle$ with a separable reduced state ρ_{BC}. When ρ_{AB} satisfies the condition $(6')$, it also satisfies the condition (1).*

Proof. Due to separability, ρ_{BC} can be written by $\rho_{BC} = \sum_i p_i |\phi_i^B, \phi_i^C\rangle\langle\phi_i^B, \phi_i^C|$. We introduce the new system $\mathcal{H}_D$ with the orthogonal basis e_i^D and the tripartite extension $\rho_{BCD} := \sum_i p_i |\phi_i^B, \phi_i^C, e_i^D\rangle\langle\phi_i^B, \phi_i^C, e_i^D|$. The monotonicity of the relative entropy $D(\rho\|\sigma) := \mathrm{Tr}(\rho\log\rho - \rho\log\sigma)$ implies that:

$$0 = H(\rho_C) - H(\rho_{BC}) = D(\rho_{BC}\|I_B \otimes \rho_C)$$

$$\leq D(\rho_{BCD}\|I_B \otimes \rho_{CD}) = \sum_i p_i(\log p_i - \log p_i) = 0\,,$$

where the first equality is from Condition (6′). So the equality holds in the above inequality. According to Petz's condition,[30,31] the channel $\Lambda_C : \mathcal{H}_C \mapsto \mathcal{H}_C \otimes \mathcal{H}_D$ with the form $\Lambda_C(\sigma) := \rho_{CD}^{1/2}((\rho_C^{-1/2}\sigma\rho_C^{-1/2}) \otimes I_D)\rho_{CD}^{1/2}$ satisfies $id_B \otimes \Lambda_C(\rho_{BC}) = \rho_{BCD}$. We introduce the system $\mathcal{H}_E$ as the environment system of Λ_C and the isometry $U : \mathcal{H}_C \mapsto \mathcal{H}_C \otimes \mathcal{H}_D \otimes \mathcal{H}_E$ as the Stinespring extension of Λ_C. So the state $|\Phi_{ABCDE}\rangle := I_{AB} \otimes U|\Psi_{ABC}\rangle$ satisfies $\rho_{BCD} = \mathrm{Tr}_{AE}|\Phi_{ABCDE}\rangle\langle\Phi_{ABCDE}|$. By using an orthogonal basis $\{e_i^{AE}\}$ on $\mathcal{H}_A \otimes \mathcal{H}_E$ we can write up the state $|\Phi_{ABCDE}\rangle = \sum_i \sqrt{p_i}|\phi_i^B, \phi_i^C, e_i^D, e_i^{AE}\rangle$. Then, the state $\rho_{AB} = \mathrm{Tr}_{CDE}|\Phi_{ABCDE}\rangle\langle\Phi_{ABCDE}| = \sum_i p_i|\phi_i^B\rangle\langle\phi_i^B| \otimes \mathrm{Tr}_E|e_i^{AE}\rangle\langle e_i^{AE}|$ is separable. This completes the proof. $\qquad\square$

Due to Lemma 6.1 and the equivalence of conditions (5′) and (6′), it suffices to show that when ρ_{BC} is nondistillable and ρ_{AB} satisfies (5′), ρ_{BC} is separable. From (5′) for ρ_{AB}, it holds that $\mathrm{rank}\,\rho_{BC} = d_A = \mathrm{rank}\,\rho_{AB} = d_C$. It follows from [Ref. 19 Theorem 10] that ρ_{BC} has to be PPT. So ρ_{BC} is separable by Lemma 3.1, and we have Theorem 2.1.

References

1. V. Coffman, J. Kundu and W. K. Wootters, *Phys. Rev. A* **61**, 052306 (2000).
2. V. Scarani and N. Gisin, *Phys. Rev. Lett.* **87**, 117901 (2001).
3. M. Koashi and A. Winter, *Phys. Rev. A* **69**, 022309 (2004).
4. T. Tsurumaru and M. Hayashi, quant-ph/1101.0064 (2011).
5. R. F. Werner, *Phys. Rev. A* **40**, 4277 (1989).
6. M. Horodecki, P. Horodecki and R. Horodecki, *Phys. Lett. A* **223**, 1 (1996).
7. M. Horodecki, P. Horodecki and R. Horodecki, *Phys. Rev. Lett.* **80**, 5239 (1998).
8. M. Horodecki and P. Horodecki, *Phys. Rev. A* **59**, 4206 (1999).
9. T. Hiroshima, *Phys. Rev. Lett.* **91**, 057902 (2003).
10. M. Hayashi and L. Chen, *Phys. Rev. A* **84**, 012325 (2011).
11. A. V. Thapliyal, *Phys. Rev. A* **59**, 3336 (1999).
12. M. P. Hertzberg and F. Wilczek, *Phys. Rev. Lett.* **106**, 050404 (2011).
13. E. M. Rains, *Phys. Rev. A* **60**, 179 (1999).
14. W. Dur, G. Vidal and J. I. Cirac, *Phys. Rev. A* **62**, 062314 (2000).
15. E. Chitambar, R. Duan and Y. Shi, *Phys. Rev. Lett.* **101**, 140502 (2008).
16. L. Chen *et al.*, *Phys. Rev. Lett.* **105**, 200501 (2010).
17. D. P. DiVincenzo *et al.*, *Phys. Rev. A* **61**, 062312 (2000).
18. P. Horodecki *et al.*, *Phys. Rev. A* **62**, 032310 (2000).
19. L. Chen and D. Z. Djokovic, *J. Phys. A: Math. Theor.* **44**, 285303 (2011).
20. T. Eggeling *et al.*, *Phys. Rev. Lett.* **87**, 257902 (2001).

21. N. J. Cerf, C. Adami and R. M. Gingrich, *Phys. Rev. A* **60**, 898 (1999).
22. C. Sabin and G. G. Alcaine, *Eur. Phys. J. D* **48**, 435 (2008).
23. J. Eisert and H. J. Briegel, *Phys. Rev. A* **64**, 022306 (2001).
24. R. Duan *et al.*, *IEEE Trans. Inform. Theory* **55**, 1320 (2009).
25. A. Peres, *Phys. Rev. Lett.* **77**, 1413 (1996).
26. M. Horodecki, P. Horodecki and R. Horodecki, *Phys. Rev. Lett.* **78**, 574 (1997).
27. P. Horodecki, J. A. Smolin, B. M. Terhal and A. V. Thapliyal, quant-ph/9910122 (1999).
28. W. K. Wootters and W. H. Zurek, *Nature* **299**, (1982).
29. M. Horodecki, J. Oppenheim and A. Winter, *Nature* **436**, 673 (2005).
30. D. Petz, *Commun. Math. Phys.* **105**, 123 (1986).
31. D. Petz, *Quart. J. Math. Oxford Ser. 2* **39**, 97 (1988).

Chapter 9

REDUCED DENSITY MATRIX AND ENTANGLEMENT ENTROPY OF PERMUTATIONALLY INVARIANT QUANTUM MANY-BODY SYSTEMS

VLADISLAV POPKOV*,† and MARIO SALERNO*

*Dipartimento di Fisica "E.R. Caianiello", Università di Salerno,
via ponte don Melillo, 84084 Fisciano (SA), Italy
†Dipartimento di Fisica, Università di Firenze,
via Sansone 1, 50019 Sesto Fiorentino (FI), Italy

In this paper we discuss the properties of the reduced density matrix of quantum many body systems with permutational symmetry and present basic quantification of the entanglement in terms of the von Neumann (VNE), Renyi and Tsallis entropies. In particular, we show, on the specific example of the spin 1/2 Heisenberg model, how the RDM acquires a block diagonal form with respect to the quantum number k fixing the polarization in the subsystem conservation of S_z and with respect to the irreducible representations of the $\mathbf{S}_n$ group. Analytical expression for the RDM elements and for the RDM spectrum are derived for states of arbitrary permutational symmetry and for arbitrary polarizations. The temperature dependence and scaling of the VNE across a finite temperature phase transition is discussed and the RDM moments and the Rényi and Tsallis entropies calculated both for symmetric ground states of the Heisenberg chain and for maximally mixed states.

Keywords: Reduced density matrix; permutation group; exact results; exact spectrum; finite temperature; Renyi entropy; von Neumann entropy; Curie–Weiss phase transition.

1. Introduction

Entanglement properties of interacting quantum many-body systems[1] lies at the heart of many quantum information processes such as measurement based quantum computation, teleportation, security of quantum key distribution protocols, super-dense coding, etc.[3] Being a principal resource for quantum information, one is interested to know how much entanglement is present in a system and how much of it can be used or created. Entanglement also provides benchmarks for success of quantum experiments. Entanglement properties are presently investigated for several spin chains,[4–12] for strongly correlated fermions[7,13,14] and pairing models,[15–17] for itinerant bosons,[18] etc.

The calculation of the entanglement involves the knowledge of the *reduced density matrix* (RDM) characterizing quantum systems in contact with the environment such as a thermal bath or a larger system of which the original system is a subsystem. In particular, the spectrum of the RDM, which by definition is real and nonnegative with all eigenvalues summing up to one, provides an intrinsic characterization of a subsystem. The relative importance of a subsystem state, indeed, is directly related to the weight that the corresponding eigenvalue has in the RDM spectrum. Thus, for instance, the fact that the eigenvalues λ_i of the RDM for a one dimensional quantum interacting subsystems decay exponentially with i implies that the properties of the subsystems are determined by only a few states. This property is crucial for the success of the density-matrix renormalization group (DMRG) method[19,20] in one dimension. In two dimensions this property is lost[21] and the DMRG method fails.

For a subsystem consisting of n sites (or n q-bits) the RDM is of rank 2^n so that for large n the calculation of the spectrum becomes a problem of exponential difficulty. While the spectrum of the full RDM for subsystems with a small number of sites (e.g., $n \leq 6$) has been calculated,[22] the full RDM for arbitrary n is, to our knowledge, exactly known only for the very special case of noninteracting quantum systems such as free fermions (see e.g., Ref. 23) or free bosons.

The aim of the present paper is to analytically calculate the elements of the RDM of permutational invariant quantum systems of arbitrary size L, for arbitrary permutational symmetry of the state of the system (labeled by an integer number $0 < r < L/2$) and arbitrary sizes n (number of q-bits) of the subsystem. We remark that the invariance under the permutational group physically implies that the interactions among sites have infinite range. As an example of such system we consider the Heisenberg model of spin $1/2$ on a full graph consisting of L sites, with fixed value of spin polarization $S_z = L/2 - N$. For this system we calculate the RDM and the entanglement von Neumann entropy (VNE) for a subsystem of arbitrary $n \geq 1$ sites for arbitrary L, N. The temperature dependence and the scaling properties of the VNE across a finite temperature phase transition occurring in the system are also discussed, and the RDM moments and the Rényi and Tsallis entropies calculated both for symmetric ground states of the Heisenberg chain and for maximally mixed states.

The plan of the paper is the following. In Sec. 1 we discuss model equations and provide basic definitions. In Sec. 2 we consider the main properties of RDM elements and show how the symmetry properties of the system allow to decompose the RDM into a block diagonal form. In Sec. 4 we present an exact analytical expression of the RDM matrix elements for arbitrary parameters values whose rigorous proof is provided in the thermodynamic limit in the Appendix B and for the case of fully symmetric states in the Appendix A. In Sec. 5 we provide an analytical characterization of the RDM spectrum and discuss scaling properties and temperature dependence of the VNE. In Sec. 6, moments of the RDM are discussed and several quantities of interest like mutual information, Renyi and

Tsallis entropies, are calculated. In the last section the main results of the paper are briefly summarized.

2. Model Equation and Basic Definitions

We consider a permutational invariant system of L spins $1/2$ on a complete graph with fixed total spin polarization $S_z = L/2 - N$ and described by the Hamiltonian

$$H = -\frac{J}{2L} \left(\mathbf{S}^2 - \frac{L}{2} \left(\frac{L}{2} + 1 \right) \right) + h S_z . \tag{1}$$

Here $\mathbf{S} \equiv (S_x, S_y, S_z)$, $S_\alpha = (1/2) \sum_{i=1}^{L} \sigma_i^\alpha$, with σ_i^α Pauli matrices acting on the factorized $\prod_1^L \otimes C_2$ space. This Hamiltonian is invariant under the action of the symmetric group $\mathbf{S}_L$[24] and conserves S_z, $[H, S_z] = 0$. A complete set of eigenstates of H are states $|\Psi_{L,N,r}\rangle$ associated to filled Young tableaux (YT) of type $\{L - r, r\}_{(N)}$ (see Ref. 25 for details), where the subscript N denotes the number of quanta present in the tableau and the symbol $\{L - r, r\}$ refers to a tableau of only two rows, with $L - r$ boxes (sites) in the first row and r in the second row. For the Hamiltonian in (1) we have:

$$H|\Psi_{L,N,r}\rangle = E_{L,N,r}|\Psi_{L,N,r}\rangle ,$$

$$E_{L,N,r} = \frac{1}{2} \left(\frac{Jr}{L}(L - r + 1) + h(L - 2N) \right) , \tag{2}$$

$$S_z|\Psi_{L,N,r}\rangle = \left(\frac{L}{2} - N \right) |\Psi_{L,N,r}\rangle$$

where $N = 0, 1, \ldots, L$ determines possible values of the spin polarization and r takes values $r = 0, 1, \ldots, \max(N, L - N)$. Notice that, due to the symmetry and antisymmetry of a YT with respect to rows and columns, respectively, the state $|\Psi_{L,N,r}\rangle$ can exist only if $N \geq r$ (for the explicit form of the state $|\Psi_{L,N,r}\rangle$ see Eq. (B.3) below). The degeneracies of the eigenvalues $E_{L,N,r}$ are given by the dimension of the corresponding YTs:

$$\deg_{L,r} = \binom{L}{r} - \binom{L}{r - 1.} \tag{3}$$

Consider a set of vectors $|\Psi_u\rangle$, $u = 1, \ldots, \deg_{L,r}$, forming an orthonormal basis in the eigenspace of H with eigenvalue $E_{L,N,r}$. We define the density matrix of the whole system as:

$$\sigma_{L,N,r} = \frac{1}{\deg_{L,r}} \sum_{u=1}^{\deg_{L,r}} |\Psi_u\rangle\langle\Psi_u| . \tag{4}$$

It can be easily shown that $\sigma_{L,N,r}$ possess the following properties:

(i) The matrix $\sigma_{L,N,r}$ has eigenvalues $\lambda_1 = \lambda_2 = \cdots = \lambda_{\deg_{L,r}} = (\deg_{L,r})^{-1}$, with remaining $2^L - \deg_{L,r}$ eigenvalues all equal to zero. This follows from the fact that each vector $|\Psi_u\rangle$ is an eigenvector of $\sigma_{L,N,r}$ with eigenvalue $1/\deg_{L,r}$. Since the spectrum of $\sigma_{L,N,r}$ is real and nonnegative with all eigenvalues summing up to 1, the remaining $2^L - \deg_{L,r}$ eigenvalues must vanish.

(ii) Matrix $\sigma_{L,N,r}$ satisfies: $(\sigma_{L,N,r})^2 = (1/\deg_{L,r})\sigma_{L,N,r}$. This follows from the definition (4) and the orthonormality condition $\langle \Psi_w | \Psi_u \rangle = \delta_{uw}$.

(iii) Matrices $\sigma_{L,N,r}$ commute with each other $[\sigma_{L,N,r}, \sigma_{L,N',r'}] = 0$. This follows from orthogonality of eigenspaces of H for different eigenvalues.

(iv) Introducing the operator P_{ij}, permuting subspaces i and j of the Hilbert space $\prod_1^L \otimes C_2$ on which the matrix $\sigma_{L,N,r}$ acts, we have that: $[\sigma, P_{ij}] = 0$ for any i, j.

This last property can be proved by considering

$$P_{ij}\sigma_{L,N,r}P_{ij} = \frac{1}{\deg_{L,r}} \sum_{u=1}^{\deg_{L,r}} P_{ij}|\Psi_u\rangle\langle\Psi_u|P_{ij}$$

$$= \frac{1}{\deg_{L,r}} \sum_{u=1}^{\deg_{L,r}} |\Psi_u'\rangle\langle\Psi_u'| \,. \tag{5}$$

The vectors $|\Psi_u'\rangle = P_{ij}|\Psi_u\rangle$ form an orthonormal basis, being $\langle \Psi_w' | \Psi_u' \rangle = \langle \Psi_w | P_{ij}^T P_{ij} | \Psi_u \rangle = \langle \Psi_w | \Psi_u \rangle = \delta_{uw}$, because $P_{ij}^T = P_{ij}$, and $(P_{ij})^2 = I$. Now, the sum $\sum_{u=1}^{\deg_{L,r}} |\Psi_u\rangle\langle\Psi_u| = I_{\deg_{L,r}}$ is a unity operator in a factor space of dimension $\deg_{L,r}$, and therefore it does not depend on the choice of the basis. Note that vector $|\Psi_u'\rangle$ belongs to the same factor space as $|\Psi_u\rangle$, because permutation P_{ij} only results in different enumeration. Consequently,

$$P_{ij}\sigma P_{ij} = \rho\,, \quad \text{or} \quad [\sigma, P_{ij}] = 0\,. \tag{6}$$

The latter property implies that in Eq. (4) the sum over the orthogonalized set of basis vector in (4) can be replaced by the symmetrization of the density matrix directly, namely $\sigma_{L,N,r} = (1/L!)\sum_P |\Psi_{12\cdots L}\rangle\langle\Psi_{12\cdots L}|$, where the sum is over all $L!$ permutations of indexes $1, 2, \ldots, L$, and $|\Psi_{12\cdots L}\rangle$ is some unit eigenvector of H with eigenvalue $E_{L,N,r}$. In particular, it is convenient to choose $|\Psi_{12\cdots L}\rangle \equiv |\Psi_{L,N,r}\rangle$,

$$\sigma_{L,N,r} = \frac{1}{L!}\sum_P |\Psi_{L,N,r}\rangle\langle\Psi_{L,N,r}|\,. \tag{7}$$

It is evident that such a sum is invariant with respect to permutations and that $\sigma_{L,N,r}$ is properly normalized:

$$\text{Tr}\,\sigma_{L,N,r} = 1\,.$$

The RDM of a subsystem of n sites is defined by tracing out $L - n$ degrees of freedom from the density matrix of the whole system:

$$\rho_{(n)} = \text{Tr}_{L-n}\sigma_{L,N,r}\,. \tag{8}$$

Due to the properties (6) and (8), $\rho_{(n)}$ does not depend on the particular choice of the n sites, and satisfies the property (6) in its subspace (we omit the explicit dependence of $\rho_{(n)}$ on L, N, r for brevity of notations).

3. RDM Properties and Block Diagonal Form

The RDM can be calculated in the natural basis by using its definition in terms of observables: $\langle \hat{f} \rangle = \text{Tr}(\rho_{(n)} \hat{f})$ where $\hat{f}$ is a physical operator acting on the Hilbert space of the $2^n \times 2^n$ subsystem. The knowledge of the full set of observables determines the RDM uniquely. Indeed, if we introduce the natural basis in the Hilbert space of the subsystem, $\prod_{k=1}^{n} \otimes C_2$, the elements of the RDM in this basis are:

$$\rho_{j_1 j_2 \cdots j_n}^{i_1 i_2 \cdots i_n} = \langle \hat{e}_{j_1 j_2 \cdots j_n}^{i_1 i_2 \cdots i_n} \rangle = \text{Tr}(\rho_{(n)} \hat{e}_{j_1 j_2 \cdots j_n}^{i_1 i_2 \cdots i_n}), \tag{9}$$

with $\hat{e}_{j_1 j_2 \cdots j_n}^{i_1 i_2 \cdots i_n} = \prod_{k=1}^{n} \otimes \hat{e}_{j_k}^{i_k}$ and $\hat{e}_j^i$ a 2×2 matrix with elements $(\hat{e}_j^i)_{kl} = \delta_{ik} \delta_{jl}$. The matrix $\hat{e}_{j_1 j_2 \cdots j_n}^{i_1 i_2 \cdots i_n}$ has only one nonzero element, equal to 1, at the crossing of the row $2^{n-1} i_1 + 2^{n-2} i_2 + \cdots + i_n + 1$ and the column $2^{n-1} j_1 + 2^{n-2} j_2 + \cdots + j_n + 1$ (all indices i, j take binary values $i_k = 0, 1$ and $j_k = 0, 1$). To determine all the RDM elements one must find a complete set of observables and compute the averages $\langle \hat{e}_{j_1 j_2 \cdots j_n}^{i_1 i_2 \cdots i_n} \rangle$. Note that a generic property of the RDM elements, which follows directly from (6), is that any permutation between pairs of indices (i_m, j_m) and (i_k, j_k) does not change its value, e.g.,

$$\rho_{j_1 j_2 \cdots j_n}^{i_1 i_2 \cdots i_n} = \rho_{j_2 j_1 \cdots j_n}^{i_2 i_1 \cdots i_n} = \rho_{j_n j_1 \cdots j_2}^{i_n i_1 \cdots i_2} = \cdots = \rho_{j_n j_{n-1} \cdots j_1}^{i_n i_{n-1} \cdots i_1}. \tag{10}$$

Another property of the RDM follows from the S_z invariance

$$\rho_{j_1 j_2 \cdots j_n}^{i_1 i_2 \cdots i_n} = 0, \quad \text{if } i_1 + i_2 + \cdots + i_n \neq j_1 + j_2 + \cdots + j_n. \tag{11}$$

Thus, for instance, the RDM for $n = 2$ has only six nonzero (four different) elements, $\rho_{00}^{00}, \rho_{01}^{10} = \rho_{10}^{01}, \rho_{10}^{10} = \rho_{01}^{01}$ and ρ_{11}^{11}, subject to normalization $\text{Tr} \, \rho_{(2)} = \rho_{00}^{00} + \rho_{10}^{10} + \rho_{01}^{01} + \rho_{11}^{11} = 1$. It is convenient to introduce the operators

$$\hat{e}_0^1 = \begin{pmatrix} 0 & 0 \\ 1 & 0 \end{pmatrix} \equiv \sigma^-, \quad \hat{e}_1^0 = \begin{pmatrix} 0 & 1 \\ 0 & 0 \end{pmatrix} \equiv \sigma^+,$$

$$\hat{e}_0^0 = \begin{pmatrix} 1 & 0 \\ 0 & 0 \end{pmatrix} \equiv \hat{p}, \quad \hat{e}_1^1 = \begin{pmatrix} 0 & 0 \\ 0 & 1 \end{pmatrix} \equiv \hat{h}.$$

If we represent a site spin up with the vector $\binom{1}{0}$ and a site spin down with the vector $\binom{0}{1}$ then $\hat{p}_k$ and $\hat{h}_k$ are spin up and spin down number operators on site k, while σ^-, σ^+ represent spin lowering and rising operators, respectively. Thus, for instance, the observable $\langle \hat{p}_1 \hat{p}_2 \hat{h}_3 \hat{h}_4 \cdots \hat{h}_n \rangle = \rho_{0011 \cdots 1}^{0011 \cdots 1}$ gives the probability to find spins down at sites $3, 4, \ldots, n$, and spins up at sites $1, 2$, while the observable $\langle \sigma_1^+ \sigma_2^+ \sigma_3^- \sigma_4^- \hat{h}_5 \cdots \hat{h}_n \rangle = \rho_{11001 \cdots 1}^{00111 \cdots 1}$ gives the probability to find spins down at sites $5, 6, \ldots, n$, spin lowering at sites $3, 4$ and spin rising at sites $1, 2$. Note that the latter operator conserves the total spin polarization since the number of lowering and rising operators is the same. Also note that the correlation functions with a nonconserved polarization vanish, e.g.,

$$\langle \sigma_1^+ \sigma_2^+ \sigma_3^- \hat{h}_4 \hat{h}_5 \cdots \hat{h}_n \rangle = \rho_{11011\cdots1}^{00111\cdots1} = 0, \tag{12}$$

in accordance with (11).

One can take advantage of the S_z invariance [e.g., Eq. (11)] to block diagonalize the RDM into independent blocks B_k of fixed polarization $k = i_1 + i_2 + \cdots + i_n = j_1 + j_2 + \cdots + j_n$ (here $k = 0, \ldots, n$ gives the number of spin up present in the subsystem). In Fig. 1 the blocks B_k appearing in the RDM $\rho_{(5)}$ have been shown for the case $n = 5$, $L = 18$, $N = 8$, $r = 6$. We remark that the $n+1$ diagonal blocks correspond to the values $s_z = (n - 2k)/2$, $k = 0, 1, \ldots, n$ the subsystem polarization can assume, being the block decomposition a direct consequence of the S_z symmetry. The dimension of the block B_k coincides therefore with the number of possible configurations that k spin up can assume on n sites, e.g., $\dim B_k = \binom{n}{k}$. One can check that the sum of the dimensions of all blocks gives the full RDM dimension, i.e., $\sum_k \dim B_k = 2^n$. Notice that the block diagonal form in the natural basis is achieved only after a number of permutations of rows and columns of the RDM have been performed. We also remark that the fact that the middle block B_3 in Fig. 1 has all vanishing anti-diagonal elements is purely accidental (see also remark at the end of Sec. 2).

Blocks B_k consist of elements $e_{j_1}^{i_1} \otimes e_{j_2}^{i_2} \otimes e_{j_3}^{i_3} \otimes \cdots \otimes e_{j_n}^{i_n}$ of the original matrix, with $\sum_1^n i_p = \sum_1^n j_p = k$ and $i_p = 0, 1$, $j_p = 0, 1$. In its turn, all elements $e_{j_1}^{i_1} \otimes e_{j_2}^{i_2} \otimes e_{j_3}^{i_3} \otimes \cdots \otimes e_{j_n}^{i_n}$ of the block B_k can be further block diagonalized (see further). In the natural basis, this diagonalization is done according to the number of pairs of type $(e_1^0 \otimes e_0^1)$ present in the elements. In the following we denote by G_Z the part

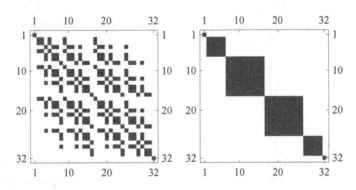

Fig. 1.   Left panel. RDM $\rho_{(5)}$ in the natural basis. Parameters are $L = 18, N = 8, r = 6$. Black boxes denote nonzero elements. Right panel. The same matrix of the left panel after the following chain of permutations of rows and columns have been applied to it: $R_{8,14} R_{14,18} R_{8,21} R_{3,29} R_{2,28} R_{6,16} R_{4,9}$. ($R_{i,j}$ denote the operator which exchange first columns i and j and then rows i and j). Black boxes denote nonzero elements. Block diagonal structure associated with values of $k = 0, 1, \ldots, 5$ is evident. The single element present in blocks $k = 0, 5$, are 1/34 and 1/153, respectively. Elements values inside other k-blocks are given in Fig. 2.

of the block associated to elements with Z pairs $(e_1^0 \otimes e_0^1)$ in it. The sub-block G_0 of the block B_k is formed by the elements containing k terms e_1^1 and $(n-k)$ terms e_0^0 in the product, i.e., $e_1^1 \otimes \cdots e_1^1 \otimes e_0^0 \otimes \cdots \otimes e_0^0$ and all permutations. All such elements lie on the diagonal, and vice versa, each diagonal element of B_k belongs to G_0. Consequently, the sub-block G_0 consists of $\binom{n}{k}$ elements. The number of elements, $\deg G_1(k)$, in the sub-block G_1 is equal to the number of elements of the type $e_1^0 \otimes e_0^1 \otimes e_{i_1}^{i_1} \otimes e_{i_2}^{i_2} \otimes \cdots \otimes e_{i_{n-2}}^{i_{n-2}}$, such that $1 + 0 + i_1 + i_2 + \cdots + i_{n-2} = k$. Using elementary combinatorics we obtain:

$$\deg G_1(k) = \binom{2}{1}\binom{n}{2}\binom{n-2}{k-1}. \tag{13}$$

Analogous calculations for arbitrary sub-block G_Z yields:

$$\deg G_Z(k) = \binom{2Z}{Z}\binom{n}{2Z}\binom{n-2Z}{k-Z}. \tag{14}$$

From the restriction $\sum_1^n i_p = \sum_1^n j_p = k$ we deduce that the block B_k contains nonempty parts $G_0, G_1, \ldots, G_{\min(k,n-k)}$, leading to the following decomposition:

$$B_k = \bigcup_{Z=0}^{\min(k,n-k)} G_Z. \tag{15}$$

Indeed, the normalization condition following from (15), gives

$$\sum_{Z=0}^{\min(k,n-k)} \deg G_Z(k) = \binom{n}{k^2}. \tag{16}$$

It is important to note that *all elements of G_Z are equal*. This is a direct consequence of the property in Eq. (10). A graphical representation of the B_k block for a particular choice of L, N, r, is given in Fig. 2.

It is instructive to discuss the structure of blocks B_k in terms of the matrices σ in (7) since this structure is directly connected with the block diagonalization of the RDM with respect the the irreps of S_n. We have, indeed, that each block B_k can be decomposed in the form:

$$B_k = \sum_{s=0}^{\min(k,n-k)} \alpha_{n,k,s}^{L,N,r} \sigma_{n,k,s} \tag{17}$$

where $\sigma_{n,k,s}$ are associated to filled YTs of type $\{n-s,s\}_{(k)}$ and $\alpha_{n,k,s}^{L,N,r}$ are coefficients related to the corresponding eigenvalues $\lambda_{n,k,s}^{L,N,r}$ of the RDM by:

$$\alpha_{n,k,s}^{L,N,r} = \lambda_{n,k,s}^{L,N,r} \deg_{n,s}. \tag{18}$$

Notice that the matrices $\sigma_{n,k,s}$ in the natural basis have dimension 2^n and coincide with the ones given in (4). In the proper basis (e.g., that of the irrep of $\mathbf{S}_n$) they have dimension $\deg_{n,s} \times \deg_{n,s}$ and contribute to B_k with a sub-block of dimension $\deg_{n,s}$ corresponding to the filled YT of type $\{n-s,s\}_{(k)}$. In performing

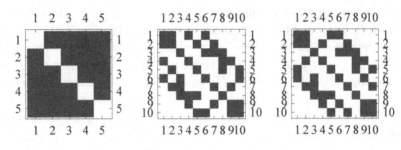

Fig. 2.   Blocks B_k for $k = 1$, 4 (left panel), $k = 2$ (central panel) and $k = 3$ (right panel) corresponding to the block diagonal form of $\rho(5)$ in Fig. 1. In the left panel white and black boxes denote elements 5/306 and 1/2448 for $k = 1$ and elements 2/51 and 1/1020 for case $k = 4$. In the center panel white boxes denote the element 1/1020 while black boxes, except for the ones along the diagonal which are all equal to 2/51, denote the element $-1/1190$. In the right panel white boxes denote the element 1/1360 while black boxes denote the element $-3/4760$, except the ones along the diagonal which are equal to 1/34. The matrices $\sigma_{n,k,s}$ involved in the decomposition in (19) have the same shapes as blocks B_k in the left panel for $\sigma_{511} = \sigma_{541}$, with white and black boxes denoting elements 1/5 and $-1/20$, respectively. The shape of matrix σ_{521} (resp. σ_{522}) is the same as in the central panel, with white boxes denoting 1/60 (resp. $-1/30$) and black ones $-1/15$ (resp. 1/30), except the ones along the diagonal which are 1/10, and shape of σ_{531} (resp. σ_{532}) is the shape is as in right panel with white box denoting 1/60 (resp. $-1/30$) and black ones $-1/15$ (resp. 1/30) except for the ones along the diagonal which are given by 1/10. The 5×5 matrices $\sigma_{510} = \sigma_{540}$, and 10×10 matrices $\sigma_{520} = \sigma_{530}$ have all elements equal to 1/5 and 1/10, respectively, while $\sigma_{500} = \sigma_{550} = 1$.

this reduction one actually achieves the diagonalization of the block B_k, as it is evident from (17) (recall that $\sigma_{n,k,s}$ have eigenvalues $1/\deg_{n,s}$). A first reduction of the matrices is achieved by accounting for the S_z symmetry discussed before, this leading to matrices $\sigma_{n,k,s}$ of size $\binom{n}{k} \times \binom{n}{k}$. In Fig. 2 are also given the matrices $\sigma_{n,k,s}$ appearing in the decomposition of blocks B_k for the specific example considered in Fig. 1 and for $k = 0, \ldots, 5$, $s = 0, 1, 2$. One can check that these matrices satisfy all properties of matrix σ given above and in particular, the number of their nonzero eigenvalues (all equal to $1/\deg_{n,s}$) coincides with the dimension $\deg_{n,s}$ of the YT to which they are associated. This implies that they can be further reduced from $\binom{n}{k} \times \binom{n}{k}$ to $\deg_{n,s} \times \deg_{n,s}$ size by eliminating the spurious $\binom{n}{k} - \deg_{n,s}$ zero eigenvalues (these eigenvalues arise because in the natural basis the dimension of the representation is larger than the one of the S_n irreps). This is achieved by using the singular valued decomposition of the matrix σ to write it in the form: $\sigma = UWV^T$, where W a diagonal matrix whose elements are the singular values and U and V are orthogonal matrices: $U^T U = V^T V = 1$, with superscript T denoting the transpose (this decomposition can be obtained very efficiently numerically[26]).

The reduction to the sub-blocks of B_k in the proper S_n representation is then achieved as: $\sigma = uwv^T$ where u and v are rectangular matrices of dimension $\binom{n}{k} \times \deg_{n,s}$ obtained from U and V by omitting the columns corresponding to the zero eigenvalues and the matrix w is a $\deg_{n,s} \times \deg_{n,s}$ diagonal matrix with the nonzero eigenvalues along the diagonal (in our case, since the nonzero eigenvalues of σ are all equal to one, w reduces to an unit matrix). The matrix w then provides the representation of $\sigma_{n,k,s}$ in the proper $\mathbf{S}_n$ space leading to the full diagonalization of the block B_k.

Thus, for example, the block diagonal form of the RDM in the right panels of Fig. 1 (see also Fig. 2) is expressed in terms of matrices σ as

$$B_1 = \lambda_{510}\sigma_{510} \oplus 4\lambda_{511} = \lambda_{540}\sigma_{540} \oplus 4\,\lambda_{541}\sigma_{541} = B_4\,, \tag{19}$$

$$B_2 = \lambda_{520}\sigma_{520} \oplus 4\,\lambda_{521}\sigma_{521} \oplus 5 \times \lambda_{522}\sigma_{522}\,, \tag{20}$$
$$B_3 = \lambda_{530}\sigma_{530} \oplus 4\,\lambda_{531}\sigma_{531} \oplus 5 \times \lambda_{532}\sigma_{532}\,,$$

with blocks $B_0 = \lambda_{500} = 1/34$, $B_5 = \lambda_{550} = 1/153$ and eigenvalues $\lambda_{510} = 11/612$, $\lambda_{520} = 76/1785$, $\lambda_{530} = 19/595$, $\lambda_{540} = 11/255$, of degeneracy one, eigenvalues $\lambda_{511} = 13/816$, $\lambda_{521} = 299/7140$, $\lambda_{531} = 299/9520$, $\lambda_{541} = 13/340$, of degeneracy four, and eigenvalues $\lambda_{522} = 13/357$, $\lambda_{532} = 13/476$, of degeneracy five (having adopted the short notation $\lambda_{nks} \equiv \lambda_{n,k,s}^{L,N,r}$, the chosen parameters $L = 18$, $N = 8$, $r = 6$ are understood).

4. Analytical Expression of RDM Elements

The main analytical property of the RDM is summarized in the following statement:

Elements g_Z of a sub-block G_Z of a block B_k of the RDM (8), for arbitrary L, N, r, n, are given by:

$$g_Z = \frac{\dbinom{L-n}{N-k}}{\dbinom{L}{N}\dbinom{N}{Z}\dbinom{L-N}{Z}} \sum_{m=0}^{Z}(-1)^m \dbinom{N-r}{Z-m}\dbinom{L-N-r}{Z-m}\dbinom{r}{m}. \tag{21}$$

This expression has been derived by extrapolating exact results obtained for finite size calculations using symbolic programs and its correctness has been checked by comparing with brute force numerical calculation of the RDM up to large sizes. Notice that Eq. (21) completely defines all elements of the RDM in the natural basis. In practice, to find the element P, Q of the RDM $(\rho_{(n)})_{PQ}$ in the natural basis one must take the binary representation of numbers $P-1$ and $Q-1$ (which provide the sets of integers $\{i_p\}$ and $\{j_p\}$, respectively), find the corresponding number Z and use (21).

A proof of the statement for arbitrary L, N, n is given in Appendix A for the specific case $r = 0$ corresponding to fully symmetric states. A proof of Eq. (21) which is valid in the thermodynamical limit $L \to \infty$ is provided in Appendix B. In this respect, we remark that in the limit $L \to \infty$ Eq. (21) simplifies to:

$$g_Z = p^{n-k}(1-p)^k \eta^Z\,, \tag{22}$$

where we denote with $p = N/L$, $\mu = r/L$ and

$$\eta = \frac{(p - \mu)(1 - p - \mu)}{p(1 - p)}.$$ (23)

For a proof of Eq. (22) see Appendix B.

5. Spectral Properties of RDM and Entanglement Entropy

The existence of two representations for the block B_k of the RDM, one in terms of matrices G_Z given in Eq. (15), the other involving matrices $\sigma_{n,k,s}$ and given in Eq. (17), have been shown in Sec. 2. These representations, together with the invariance of G_Z and $\sigma_{n,k,s}$ with respect to permutations, imply the existence of linear relations of the form

$$\hat{G}_Z(k) = \sum_{s=0}^{\min(k,n-k)} \beta_Z(k,s)\sigma_{n,k,s},$$ (24)

where $\beta_Z(k,s)$ are constants and $\hat{G}_Z$ denotes the matrix formed by all elements of G_Z. Since $\sigma_{n,k,s}$ commute for different s (see the property (iii) of matrices σ in Sec. 1), we also have that $\hat{G}_Z$ commute:

$$[\hat{G}_Z(k), \hat{G}_{Z'}(k)] = 0.$$ (25)

This also implies that all RDM eigenvalues $\lambda_{n,k,s}^{L,N,r}$ must be linear combinations of elements g_Z of matrices G_Z. One can show, indeed, that the general expression of the RDM eigenvalues is:

$$\lambda_{n,k,s}^{L,N,r} = \sum_{Z=0}^{\min(k,n-k)} \alpha_Z^{(s)}(n,k)g_Z,$$ (26)

with coefficients $\alpha_Z^{(s)}(n,k)$ given by:

$$\alpha_Z^{(s)}(n,k) = (-1)^Z \sum_{i=0}^{k-s}(-1)^i \binom{k-s}{i}\binom{n-k+s}{i}\binom{s}{Z-i},$$ (27)

where Z, $s = 0, 1, \ldots, k$. From Eq. (27) one can see that $\alpha_Z^{(s)}(n,k)$ are *integer coefficients* which, due to the property (25), do not depend on the characteristics of the original state L, N and r. Thus the dependence of the RDM eigenvalues on these parameters enters only through the elements g_Z (21). Moreover, one can show that they satisfy the following relations:

$$\alpha_0^{(s)}(n,k) = 1,$$ (28)

$$\sum_{Z=0}^{\min(k,n-k)} \alpha_Z^{(0)}(n,k) = \binom{n}{k},$$ (29)

$$\sum_{Z=0}^{\min(k,n-k)} \alpha_Z^{(s)}(n,k) = 0 \quad \text{for } s > 0, \tag{30}$$

$$\sum_{Z=0}^{\min(k,n-k)} \alpha_Z^{(s)}(n,k) \binom{Z}{p} = 0, \quad p = 0,1,\dots,s-1, \tag{31}$$

$$\sum_{Z=0}^{\min(k,n-k)} \alpha_Z^{(s)}(n,k) \binom{Z}{k} = \alpha_k^{(s)}(n,k) = (-1)^s \binom{n-k-s}{k-s}. \tag{32}$$

a proof of which can be found for special cases in Ref. 29.

From Eqs. (21), (22), (26) and (27), the explicit analytical form of the *complete spectrum of the RDM* is obtained.

The knowledge of the RDM spectrum allows to investigate the bipartite entanglement, e.g., the entanglement of a subsystem of size n with respect to the rest of the system (see Refs. 1 and 2 for a review). This is done in terms of the entanglement entropy which for pure states at zero temperature coincides with the VNE

$$S_{(n)} = -\text{Tr}(\rho_{(n)} \log_2 \rho_{(n)}) = -\sum \lambda_k \log_2 \lambda_k, \tag{33}$$

where λ_k the eigenvalues of the RDM $\rho_{(n)}$, obtained from the density matrix ρ of the whole system as $\rho_{(n)} = tr_{(L-n)}\rho$. For the infinite range ferromagnetic Heisenberg model at zero temperature the density matrix of the whole system is a projector on the symmetric ground state $\rho = |\Psi(L,N)\rangle\langle\Psi(L,N)|$ considered in Ref. 11 where it was shown that $\lambda_k = \binom{k}{n}\binom{N-k}{L-n}/\binom{N}{L}$, where $k = 0,1,\dots,\min(n,N)$. In the limit of large n the VNE becomes:

$$S_{(n)} \approx \frac{1}{2}\log_2(2\pi epq) + \frac{1}{2}\log_2 \frac{n(L-n)}{L}. \tag{34}$$

One can show that a zero temperature (e.g., $\mu = r/L = 0$)[11] the spectrum of the RDM is described by a binomial distribution $\lambda_k = p^k q^{n-k} \binom{n}{k}$ which converges to a Gaussian for large n

$$\lambda_k \approx \frac{1}{\sqrt{2\pi\sigma^2}} \exp\left(-\frac{n^2\left(p - \frac{k}{n}\right)^2}{2\sigma^2}\right), \tag{35}$$

where $\sigma^2 = np(1-p) \gg 1$. For finite temperature one introduces the thermal VNE for a block of size n as:

$$S_{(n)}(\beta) = -\text{Tr}(\tilde{\rho}_{(n)}(\beta) \log_2 \tilde{\rho}_{(n)}(\beta)), \tag{36}$$

$$\tilde{\rho}_{(n)}(\beta) = \frac{1}{Z} \sum_r d_r e^{-\beta E_r} \langle \rho_{(n)}(r) \rangle, \tag{37}$$

where $\tilde{\rho}_{(n)}(\beta)$ is the thermal RDM, $\mathcal{Z}$ is the partition function and $\langle \cdot \rangle$ denotes the equal weight (thermic) average over all orthogonal degenerate states, corresponding to a given permutational symmetry. Note that $\langle \rho_{(n)}(r) \rangle$ commutes with any permutation operator and does not depend on the choice of sites in the block but only on its size n. Also note that the matrices $\langle \rho_{(n)}(r) \rangle$ commute for different r

$$[\langle \rho_{(n)}(r) \rangle, \langle \rho_{(n)}(r') \rangle] = 0, \tag{38}$$

so that the diagonalization of $\tilde{\rho}_{(n)}(\beta)$ is reduced to the diagonalization of $\langle \rho_{(n)}(r) \rangle$ for arbitrary r. From Eq. (37) the computation of the temperature-dependent VNE is easily made with the help of the general expression of the eigenvalues of the RDM in Eqs. (26) and (27) for states of arbitrary permutational symmetry. While $p = N/L$ is the system polarization, the relation between the temperature T and the parameter $\mu = r/L$ is fixed by the condition of the minimum of the free energy of the whole system defined by the spectrum (2) and its degeneracy. It has the form (see Refs. 27 and 28)

$$\frac{J}{2T} = \frac{1}{(1 - 2\mu)} \ln \left(\frac{1 - \mu}{\mu} \right). \tag{39}$$

The scaling of the thermal VNE across a phase transition, which occurs in the system with infinite range interactions at finite temperature $T_c \neq 0$,[25] has been considered in Refs. 27 and 28. In this case it was shown that the VNE of a block of size n scales as:

$$S_{(n)} = \begin{cases} \dfrac{1}{2} \log n + \dfrac{1}{2} \log 2\pi epq & \text{for } T = 0 \\[2mm] nH(\mu) + \dfrac{1}{2} \log n + C(p, \mu) & \text{for } 0 < T < T_c \\[2mm] nH(\min(p, 1 - p)) & \text{for } T \geq T_c \end{cases} \tag{40}$$

where $H(a) = -a \log a - (1 - a) \log(1 - a)$, $q = 1 - p$ and $C(p, \mu)$ does not depend on n.

Another quantity of interest strictly related to the entanglement entropy is the *mutual information*, I_{AB}, which measures the work necessary to erase all correlations in the bipartite system[1,2]:

$$I_{AB} = S(A) + S(B) - S(AB), \tag{41}$$

where $S(X)$ is the VNE of the subsystem X. At nonzero temperature we find for a subsystem of size n of a system with size L, using (40):

$$I_{AB}(n, L, T) = \frac{1}{2} \log(n(L - n)) + \text{Const.} \tag{42}$$

for all $T < T_c$ and $I_{AB}(T) = 0$ for $T \geq T_c$.

6. Moments of the Reduced Density Matrix and Rényi and Tsallis Entropies

Besides the entanglement entropy, the Rényi, $R(\alpha)^{30}$ and Tsallis $T(\alpha)^{31}$ entropies, defined as:

$$S_R(\alpha, n) = \frac{\log \mathrm{Tr}(\rho^\alpha_{(n)})}{1 - \alpha}, \quad S_T(\alpha, n) = \frac{\mathrm{Tr}(\rho^\alpha_{(n)}) - 1}{1 - \alpha}, \tag{43}$$

with α a positive real number, are also commonly used as a measure of entanglement. Notice that both expressions reduce the VNE in the limit $\alpha \to 1$. The knowledge of these generalized entropies requires the computation of $\mathrm{Tr}(\rho^\alpha_{(n)})$ which, except special cases (see below) it is a very difficult task. For α positive integers, however, the moments $\mathrm{Tr}(\rho^\alpha_{(n)})$ can be computed using a quantum field theory (QFT) procedure which is known as the *replica method*[32,33] (reminiscent of the "replica trick" of disordered systems). In this case the entanglement entropy is obtained through an analytical continuation of $\mathrm{Tr}(\rho^\alpha_{(n)})$ from positive integers to real α values, using

$$S_{(n)} = -\lim_{\alpha \to 1} \frac{\partial}{\partial \alpha} \mathrm{Tr}[\rho_{(n)}{}^\alpha]. \tag{44}$$

In the case of 1+1 conformal field theories critical models at zero temperature (for ground state) the displays universal properties, namely

$$\mathrm{Tr}(\rho^\alpha_{(n)}) = C_\alpha \left(\frac{L}{\pi} \sin \frac{\pi n}{L} \right)^{-c\frac{\alpha - 1/\alpha}{6}}, \tag{45}$$

where c is the central charge of the underlying conformal field theory. Similarly, for the quantum XY chain with periodic boundary conditions at zero temperature it has been shown that the RDM is independent on the block size n and the moments can be expressed in the form[34]

$$\left(\frac{1}{2} \mathrm{Tr}(\rho^\alpha_{(n)}) \right)^{1/2} = \frac{\prod_{m=1}^{\infty}(1 + e^{-2m\epsilon\alpha})}{\prod_{m=1}^{\infty}(1 + e^{-2m\epsilon})^\alpha}, \tag{46}$$

where ϵ depends on the anisotropy and transverse field parameters. Except these and few other cases, analytical properties of RDM for interacting systems are largely unexplored. The characterization of the RDM spectrum given for permutational invariant systems allows to provide another exact result for the RDM moments which is not accessible by QFT methods (our model is not conformal invariant).

In particular, the ground states of the ferromagnetic Heisenberg chain being characterized by the YT with $r = 0$, are fully symmetric states with respect to the permutations (see appendix). For these states then one can obtain the analytical expression of $\mathrm{Tr}(\rho^\alpha)$ straightforwardly, using the Gaussian distribution of the symmetric RDM eigenvalues derived in (35). The approximation $\sum_k \cdots \approx n \int \cdots dx$, indeed, readily provides:

$$\mathrm{Tr}(\rho^\alpha) \approx n \int_{-\infty}^{\infty} \frac{1}{(2\pi\sigma^2)^{\alpha/2}} \exp\left(-\frac{n^2(p - x)^2}{2\sigma^2}\alpha \right) dx$$

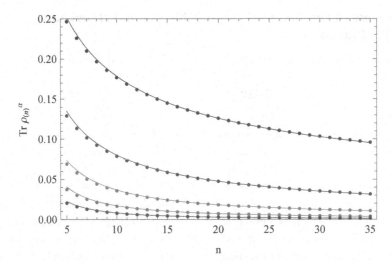

Fig. 3.   Traces of the powers $\alpha = 2, 2.5, 3, 3.5, 4$ (from top to bottom, respectively) of the RDM versus the block size n for symmetric states $r = 0$, as obtained from the analytical expression (47) for $p = 0.5$ (continuous curves) and from exact expressions of RDM eigenvalues (26) and (27) (full dots) for $L = 2 * 10^3$.

$$= \frac{1}{\sqrt{\alpha}} (2\pi p q)^{\frac{1-\alpha}{2}} n^{\frac{1-\alpha}{2}} . \tag{47}$$

In Fig. 3 we compare the behavior of the traces of the RDM powers with the block size n, as obtained from Eq. (47) and from exact expressions of RDM eigenvalues. We see that the agreement is very good this confirming the correctness of our analytical derivation.

In the case the original global state has the form of a maximally mixed state, i.e., is the sum of equally weighted projectors on symmetric states $|\Psi_{L,N}\rangle$ of the form $\rho = (1/L+1) \sum_{N=0}^{L} |\Psi_{L,N}\rangle\langle\Psi_{L,N}|$, the RDM has one eigenvalue only $\lambda_k = (1/n+1)$, which is degenerate $n+1$ times. In this case, then,

$$\mathrm{Tr}(\rho_{(n)}^{\alpha}) = (n+1)^{1-\alpha} . \tag{48}$$

Notice that the entanglement entropy at $T = 0$ in (40) follows from (47) using the expression (44) of the QFT replica method. Another quantity directly related to the RDM moments is the *effective dimension* defined as $d_{\mathrm{eff}} = 1/\mathrm{Tr}(\rho_{(n)}^2)$. Summarizing the above results, we have for this quantity that:

$$d_{\mathrm{eff}} = \begin{cases} \sim n & \text{for maximally mixed symmetric state} \\ \sim \sqrt{n} & \text{for pure symmetric state} \\ \sim n^{1/4} & \text{for critical XXZ model ground state} \\ \sim n^{c/4} & \text{for a critical state with central charge } c \end{cases}$$

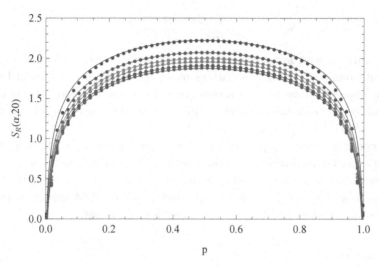

Fig. 4. Rényi entropies for symmetric ground states of the Heisenberg chain versus polarization $p = N/L$ for the cases $\alpha = 1, 2, 3, 4, 6, 8$ (curves from top to bottom, respectively) and for $n = 20$. Continuous curves refer to the analytical expression in (49) while the full dots are obtained from exact calculations using RDM eigenvalues in (26) and (27) for $L = 2 * 10^3$ and same polarization.

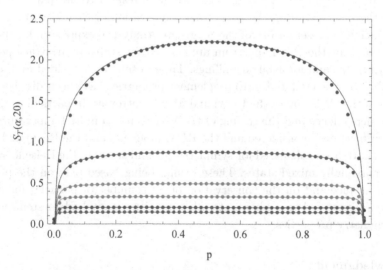

Fig. 5. Same as in Fig. 4 but for Tsallis entropy. Continuous curves refer to the analytical expression in (50) while the full dots are obtained from exact calculations using RDM eigenvalues.

From the expression of $\mathrm{Tr}(\rho_{(n)}^{\alpha})$ in (47) the Rényi and Tsallis entropies for fully symmetric states follow as:

$$S_R(\alpha, n) = \frac{1}{2} \log(2\pi npq) - \frac{\log \alpha}{2(1 - \alpha)}, \tag{49}$$

$$S_T(\alpha, n) = \frac{1}{\sqrt{\alpha}} \frac{(2\pi npq)^{\frac{1-\alpha}{2}} - 1}{1 - \alpha}. \tag{50}$$

In Figs. 4 and 5, we compare the above analytical expressions for the Rényi and Tsallis entropies with exact calculations using the RDM eigenvalues in Eqs. (26) and (27), from which we see that a very good agreement is found. Also notice that in the limit $\alpha = 1$ both entropies reduce to the entanglement entropy (40) at $T = 0$: $S_R(1, n) = S_T(1, n) = (1/2) \log 2\pi epqn$.

In general, for arbitrary permutational symmetries and for finite temperatures, one must recourse to direct calculations using the general expression (26) for the RDM eigenvalues, since it is not easy in these cases to give simple analytical expressions of α. The study of the analytical properties of the RDM moments represents an interesting problem which deserves further investigations.

7. Conclusions

To summarize, we have provided explicit analytical expression of the RDM of a subsystem of arbitrary size n of a permutational invariant quantum many body system of arbitrary size L and characterized by a state of arbitrary permutational symmetry. We have shown, on the specific example of the spin 1/2 Heisenberg model, that the RDM acquires a block diagonal form with respect to the quantum number k fixing the polarization in the subsystem conservation of S_z and with respect to the irreducible representations of the $\mathbf{S}_n$ group. Analytical expression for the RDM elements and for the RDM spectrum are derived for states of arbitrary permutational symmetry and for arbitrary fillings. These results are provided by Eqs. (21), (22) and (27) presented above. Entanglement properties have been discussed both in terms of the VNE and of the Renyi and Tsallis entropies. In particular, the temperature dependence and the scaling of the VNE across a finite temperature phase transition have been considered and the RDM moments and the Rényi and Tsallis entropies have been calculated for symmetric ground states of the Heisenberg chain and for maximally mixed states. These results being based only on the permutational invariance and on the conservation of S_z (number of particles for nonspin systems) are expected to apply also to other quantum many-body systems with the same symmetry properties.

Acknowledgments

This paper is written in honor of the 60th birthday of Professor Vladimir Korepin. V. P. thanks V. Korepin and V. Vedral for stimulating discussions, the Center for Quantum Technologies, Singapore for hospitality, and the University of Salerno for a research grant (Assegno di Ricerca no. 1508). M. S. acknowledges support from the Ministero dell' Istruzione, dell' Università e della Ricerca (MIUR) through a *Programma di Ricerca Scientifica di Rilevante Interesse Nazionale* (PRIN)-2008 initiative.

Appendix A. RDM Elements for Symmetric States

In this appendix we provide a proof of Eq. (21) which is valid for fully symmetric states (case $r = 0$) such as, for example, the ground state of the ferromagnetic Heisenberg chain. For $r = 0$, the corresponding YT is nondegenerate and the state of the full system is pure: $\rho = |\Psi_{L,N}\rangle\langle\Psi_{L,N}|$ with $|\Psi_{L,N}\rangle$ the symmetric state

$$|\Psi_{L,N}\rangle = \binom{L}{N}^{-1/2} \sum_P |\underbrace{\uparrow\uparrow\cdots\uparrow}_{N}\underbrace{\downarrow\downarrow\cdots\downarrow}_{L-N}\rangle, \tag{A.1}$$

where the sum is over all possible permutations. Since all L sites are equivalent due to permutational invariance, any choice of n sites $i_1, i_2, \ldots, i_n$ within L sites gives the same RDM, which we denote by $\rho_{(n)}^{L,N,0} = \text{Tr}_{L-n}\rho$. It has been shown in Ref. 11 that $\rho_{(n)}^{L,N,0}$ takes form:

$$\rho_{(n)}^{L,N,0} = \sum_{k=0}^{n} \frac{\binom{L-n}{N-k}}{\binom{L}{N}} |\Psi_{n,k}\rangle\langle\Psi_{n,k}|.$$

In the natural basis the matrix elements of RDM are given by the above discussed values of observables. Using (A.1), one explicitly computes all RDM elements as

$$(\rho_{(n)}^{L,N,0})_Q^P = (\rho_{(n)}^{L,N,0})_{i'j'\cdots m'}^{ij\cdots m} = \delta_{i'+j'\cdots+m'}^{i+j+\cdots+m} \frac{\binom{L-n}{N-w}}{\binom{L}{N}} \tag{A.2}$$

with $w = i+j+\cdots+m$ (the sets $ij\cdots m$ and $i'j'\cdots m'$ are binary representation of numbers $P-1, Q-1$). In this case we have that all elements of a block B_k are equal (this is not true for $r \geq 0$). We also see that the the elements in Eq. (A.2) are the same as those obtained from Eq. (21) for $r = 0$. Note that in the thermodynamic limit $\eta = 1$, and $\binom{L-n}{N-k}/\binom{L}{N} \to p^{n-k}(1-p)^k$ in agreement with Eq. (22).

Appendix B. RDM Elements in the Thermodynamic Limit

To calculate the RDM, we shall use the representation (7) for the density matrix of the whole system σ, rewritten in the form

$$\rho_{(n)} = \text{Tr}_{L-n}\left\{\frac{1}{L!}\sum_P |\Psi_{L,N,r}\rangle\langle\Psi_{L,N,r}|\right\}$$

$$= \text{Tr}_{L-n}\left\{\frac{1}{n!}\frac{1}{(L-n)!}\frac{1}{\binom{L}{n}}\sum_{P(n)}\sum_{P(L-n)}\sum_{i_1\neq i_2\neq\cdots\neq i_n} |\Psi_{L,N,r}\rangle\langle\Psi_{L,N,r}|\right\}. \tag{B.1}$$

Note that the $L!$ permutations can be done in three steps: first, choose at random n sites $i_1 \neq i_2 \neq \cdots \neq i_n$ among the L sites. There are $\binom{L}{n}$ such choices. Then, permute the chosen n sites, the total number of such permutations being $n!$. Finally, permute the remaining $L - n$ sites, the total number of such permutations being $(L - n)!$ The latter step (c) under the trace operation is irrelevant because these degrees of freedom will be traced out. The operation permuting n sites commutes with the trace operation since Tr_{L-n} does not touch the respective subset of n sites. Consequently, (B.1) can be rewritten as:

$$\rho_{(n)} = \frac{1}{n!} \sum_{P_{(n)}} \text{Tr}_{L-n} \frac{1}{\binom{L}{n}} \sum_{i_1 \neq i_2 \neq \cdots \neq i_n} |\Psi_{L,N,r}\rangle\langle\Psi_{L,N,r}| . \tag{B.2}$$

We recall here that a filled YT of type $\{L-r, r\}_{(N)}$ contains a mixed symmetry part with $2r$ sites and r spin up , and a fully symmetric part with $L - 2r$ sites and $N - r$ spin up (in the following we adopt an equivalent terminology which refers to spins up as particles and to a spins down as holes). This implies that the corresponding wavefunction $|\Psi_{L,N,r}\rangle$ factorizes into symmetric and antisymmetric parts as:

$$\Psi = |\phi_{12}\rangle \otimes |\phi_{34}\rangle \otimes \cdots \otimes |\phi_{2r-1,2r}\rangle \otimes |\Psi_{L-2r,N-r}\rangle_{2r+1,2r+2,\ldots,L} \tag{B.3}$$

with the antisymmetric part consisting of the first r factors of the type

$$|\phi_{12}\rangle = \frac{1}{\sqrt{2}}(|10\rangle_{12} - |01\rangle_{12}) \tag{B.4}$$

and with the symmetric part, $|\Psi_{L-2r,N-r}\rangle$, given by (A.1). A general property of factorized states implies that if the global wavefunction is factorized, $|\Phi\rangle = |\psi\rangle_I |\phi\rangle_{II}$ and out of n sites of the subsystem, n_1 sites belong to subset I, and the remaining $n_2 = n - n_1$ sites belong to the subset II, then the RDM factorizes as well:

$$\rho_{(n)} = \rho^I_{(n_1)} \otimes \rho^{II}_{(n_2)} . \tag{B.5}$$

To do the averaging, we note that among total number of choices $\binom{L}{n}$ there are (a) $\binom{L-2r}{n}$ possibilities to choose n sites inside the symmetric part of the tableau, containing $N - r$ particles, (b) $2r\binom{L-2r}{n-1}$ possibilities to choose $n-1$ sites inside the symmetric part of the tableau and one site in the antisymmetric part (c) $\binom{2r}{2}\binom{L-2r}{n-2}$ possibilities to choose $n - 2$ sites inside the symmetric part of the tableau and two sites in the antisymmetric part and so on. The contributions given by (a) and (b) to the right hand side of (B.2) for $\rho^{L,N,r}_{(n)}$ are given, according to (B.5), by:

$$\left\langle \binom{F}{n} \rho^{F,M,0}_{(n)} + 2r \binom{F}{n-1} \rho^{F,M,0}_{(n-1)} \otimes \rho_{\frac{1}{2}} \right\rangle , \tag{B.6}$$

with $F = L - 2r$, $M = N - r$ and with $\rho_{\frac{1}{2}} = I/2$ the density matrix corresponding to a single site in the antisymmetric part of the tableau. Brackets $\langle \cdot \rangle = (1/n!)\sum_P$ denote the average with respect to permutations of n elements. The contribution due to (c) to (B.2) splits into two parts since the $\binom{2r}{2}$ possibilities to choose two sites in the antisymmetric part of the tableau consist of $4\binom{r}{2}$ choices with two sites into different columns and the remaining r choices with both sites belonging to a same column. For the former choice, the corresponding density matrix is $\rho_{(n-2)}^{F,M,0} \otimes \rho_{\frac{1}{2}} \otimes \rho_{\frac{1}{2}}$, while for the latter case is given by $\rho_{(n-2)}^{F,M,0} \otimes \rho_{\text{asymm}}$, with

$$
\rho_{\text{asymm}} = \left| \frac{1}{\sqrt{2}}(10-01) \right\rangle \left\langle \frac{1}{\sqrt{2}}(10-01) \right| = \frac{1}{2} \begin{pmatrix} 0 & 0 & 0 & 0 \\ 0 & 1 & -1 & 0 \\ 0 & -1 & 1 & 0 \\ 0 & 0 & 0 & 0 \end{pmatrix}. \tag{B.7}
$$

Proceeding in the same manner for arbitrary partitions of Z sites in the antisymmetric part of the tableau and $n - Z$ sites in the symmetric part, we get:

$$
\rho_{(n)}^{L,N,r}\binom{L}{n} = \left\langle \sum_{Z=0}^{\min(2r,n)} \binom{F}{n-Z} \rho_{(n-Z)}^{F,M,0} \sum_{i=0}^{[Z/2]} \binom{r}{i} \left(\prod_{1}^{i} \otimes \rho_{\text{asymm}} \right) \right.
$$

$$
\left. \times 2^{Z-2i} \binom{r-i}{Z-2i} \left(\prod_{1}^{Z-2i} \otimes \rho_{\frac{1}{2}} \right) \right\rangle. \tag{B.8}
$$

From this the general scheme for the decomposition of the general RDM becomes evident. In the above formula, the products $\prod_{i}^{Q}$ with $Q < i$ are discarded. The matrix elements $\rho_{(k)}^{F,M,0}$ are given by (A.2).

For simplicity of presentation, we prove Eq. (22) for the case $Z = k$ and then outline the proof for arbitrary Z.

In the thermodynamic limit one can neglect the difference between factors like $4\binom{r}{2}$ and $\binom{2r}{2}$ in Eq. (B.8). The latter then can be then rewritten in a simpler form as:

$$
\binom{L}{n} \rho_{(n)}^{L,N,r} = \left\langle \binom{F}{n} \rho_{(n)}^{F,M,0} + \binom{2r}{1} \binom{F}{n-1} \rho_{(n-1)}^{F,M,0} \otimes \rho_{\frac{1}{2}} \right.
$$

$$
\left. + \binom{2r}{2} \binom{F}{n-2} \rho_{(n-2)}^{F,M,0} \otimes \rho_{\frac{1}{2}} \otimes \rho_{\frac{1}{2}} + \cdots \right\rangle. \tag{B.9}
$$

Note that one can omit all terms in (B.8) containing ρ_{asymm} since the respective coefficients correspond to probabilities of finding two adjacent sites in the asymmetric part of the YT (proportional to r), which vanish in the thermodynamic limit, respect to the total number of choices which is of order of r^2. A sub-block G_Z of a block k consists of all elements of the matrix $\rho_{(n)}$ having Z pairs of e_1^0, e_1^0 in its tensor representation, like e.g., $(e_1^0 \otimes e_0^1)^{\otimes Z} \otimes e_{i_1}^{i_1} \otimes e_{i_2}^{i_2} \otimes \cdots \otimes e_{i_{n-2Z}}^{i_{n-2Z}}$, such that $Z + i_1 + i_2 + \cdots + i_{n-2Z} = k$. The total number of elements $g_Z \subset G_Z$ in $\rho_{(n)}^{L,N,r}$ is equal to the number of distributions of Z objects e_1^0, Z objects e_0^1 and $(k - Z)$ objects e_1^1 on n places, given by:

$$
\deg G_Z = \frac{n!}{Z!Z!(k-Z)!(n-k-Z)!} \tag{B.10}
$$

(this is another way of writing (14)). Each term W in the sum (B.9) after averaging will acquire the factor

$$\Gamma(W) = \frac{\deg G_Z(W)}{\deg G_Z} \tag{B.11}$$

where $\deg G_Z(W)$ is a total number of g_Z elements in the term W, provided all of them are equal. For instance, $\deg G_Z(\rho^{F,M,0}_{(n)}) = \deg G_Z$, $\deg G_Z(\rho^{F,M,0}_{(n-m)} \otimes (\rho_{\frac{1}{2}})^{\otimes m}) = \binom{2Z}{Z}\binom{n-m}{2Z}\sum_{m_1=0}^{m}\binom{m}{m_1}\binom{n-2Z-m}{k-Z-m_1}$ (the last formula is only true for $k = Z$, otherwise elements constituting $G_Z(W)$ are not all equal). Restricting to the case $k = Z$ and denoting $W_m = \rho^{F,M,0}_{(n-m)} \otimes (\rho_{\frac{1}{2}})^{\otimes m}$, we have:

$$\Gamma(W_m) = \Gamma_m = \frac{\binom{n-m}{2Z}}{\binom{n}{2Z}}. \tag{B.12}$$

It is worth to note that the element $g_Z \subset G_Z$ is simply given by:

$$\binom{L}{n} g_Z = \Gamma_0 \binom{F}{n} g_0^{(n,k)} + \Gamma_1 \binom{2r}{!}\binom{F}{n-1} \frac{g_0^{(n-1,k)}}{2}$$

$$+ \Gamma_2 \binom{2r}{2}\binom{F}{n-2} \frac{g_0^{(n-2,k)}}{2^2} + \cdots \tag{B.13}$$

with $q = 1 - p$ and $g_0^{(n,k)} = (\binom{F-n}{M-k}/\binom{F}{M}) \approx (p - \mu/1 - 2\mu)^{n-k}(q - \mu/1 - 2\mu)^k$ is the element of a $\rho^{F,M,0}_{(n)}$ corresponding to a block with k particles (the factors Γ_m are due to the averaging while the factors $1/2^m$ come from $(\rho_{\frac{1}{2}})^{\otimes m}$). Restricting to the case $k = Z$, and taking into account

$$\frac{\binom{F}{n-m}}{\binom{L}{n}} \approx \frac{n!}{(n-m)!}\frac{(1-2\mu)^n}{F^m}, \quad \binom{2r}{m} \approx \frac{(2\mu)^m}{m!}L^m, \tag{B.14}$$

so that

$$\frac{\binom{F}{n-m}}{\binom{L}{n}}\binom{2r}{m}\frac{1}{2^m} \approx \binom{n}{m}\mu^m(1-2\mu)^{n-m},$$

we finally obtain, using (B.13), that:

$$g_Z = \sum_{m=0}^{n-2Z}\mu^m\binom{n-2Z}{m}(p-\mu)^{n-m-Z}(q-\mu)^Z$$

$$= (p-\mu)^{n-Z}(q-\mu)^Z\sum_{m=0}^{n-2Z}\frac{\mu^m}{(p-\mu)^m}\binom{n-2Z}{m}$$

$$= (p - \mu)^{n-Z}(q - \mu)^Z \left(\frac{p}{p - \mu} \right)^{n-2Z}$$

$$= p^{n-Z}q^Z \left(\frac{(p - \mu)(q - \mu)}{pq} \right)^Z$$

$$= p^{n-Z}q^Z\eta^Z = \eta^Z g_0, \tag{B.15}$$

with g_0 the diagonal element in the same block $k = Z$. In the last calculation we used the relation $(\binom{n}{m}\binom{n-m}{2Z})/\binom{n}{n-Z} = \binom{n-2Z}{m}$. This proves formula (22) for the particular case $k = Z$ and arbitrary n.

For arbitrary k, Z, one proceeds in similar manner as for the case $k = Z$ case. Since the respective calculations are tedious and not particularly illuminating, we omit them and give only the final result:

$$g_Z = \sum_{m=0}^{n-2Z} \mu^m \sum_{i=\max(0,Z+k-n+m)}^{\min(m,k-Z)} (p - \mu)^{n-m-k+i}$$

$$\times (q - \mu)^{k-i} \binom{k - Z}{i} \binom{n - k - Z}{m - i}, \tag{B.16}$$

which, after some algebraic manipulation, can be rewritten in the form:

$$g_Z = (p - \mu)^{n-k}(q - \mu)^k \sum_{j=0}^{n-k-Z} \left(\frac{\mu}{p - \mu} \right)^j \binom{n - k - Z}{j}$$

$$\times \sum_{i=0}^{k-Z} \left(\frac{\mu}{q - \mu} \right)^i \binom{k - Z}{i} = \eta^Z p^{n-k}q^k. \tag{B.17}$$

This concludes the proof of Eq. (21) in the thermodynamic limit $L \to \infty$.

References

1. L. Amico *et al.*, *Rev. Mod. Phys.* **80**, 517 (2008).
2. J. Cardy, *Eur. Phys. J. B* **64**, 321 (2007).
3. M. A. Nielsen and I. Chuang, *Quantum Computation and Quantum Communication* (Cambridge University Press, Cambridge, 2000).
4. G. Vidal *et al.*, *Phys. Rev. Lett.* **90**, 227902 (2003).
5. J. I. Latorre, E. Rico and G. Vidal, *Quantum Inf. Comput.* **4**, 48 (2004).
6. B. Q. Jin and V. E. Korepin, *Phys. Rev. A* **69**, 062314 (2004).
7. V. E. Korepin, *Phys. Rev. Lett.* **92**, 096402 (2004).
8. I. Peschel, *J. of Stat. Mech.: Theory Exp.* P1200 (2004).
9. I. Peschel, *J. Phys. A* **38**, 4327 (2005).
10. A. R. Its, B. Q. Jin and V. E. Korepin, *J. Phys. A* **38**, 2975 (2005).
11. V. Popkov and M. Salerno, *Phys. Rev. A* **71**, 012301 (2005).
12. V. Popkov, M. Salerno and G. Schütz, *Phys. Rev. A* **72**, 032327 (2005).
13. S. J. Gu *et al.*, *Phys. Rev. Lett.* **93**, 086402 (2004).
14. A. Anfossi *et al.*, *Phys. Rev. Lett.* **95**, 056402 (2005).
15. P. Zanardi, *Phys. Rev. A* **65**, 042101 (2002).

16. Y. Shi, *J. Phys. A* **37**, 6807 (2004).
17. H. Fan and S. Lloyd, *J. Phys. A: Math. Gen.* **38**, 5285 (2005).
18. K. Helmerson and L. You, *Phys. Rev. Lett.* **87**, 170402 (2001).
19. S. R. White, *Phys. Rev. Lett.* **69**, 2863 (1992).
20. S. R. White, *Phys. Rev. B* **48**, 10345 (1993).
21. M.-C. Chung and I. Pechel, *Phys. Rev. B* **64**, 064412 (2001).
22. B. Nienhuis, M. Campostrini and P. Calabrese, *J. Stat. Mech.* P02063 (2009).
23. I. Peschel, *J. Stat. Mech.* P06004 (2004).
24. B. E. Sagan, *The Symmetric Group* (Springer, Berlin, 1991).
25. M. Salerno, *Phys. Rev. E* **50**, 4528 (1994).
26. W. H. Press *et al.*, *Numerical Recipes* (Cambridge University Press, Cambridge, 1989).
27. V. Popkov and M. Salerno, *Europhys. Lett.* **84**, 30007 (2008).
28. M. Salerno and V. Popkov, *Phys. Rev. E* **82**, 011142 (2010).
29. M. Salerno and V. V. Popkov, *Acta Applicanda Matemathicae* **115**, 75 (2011).
30. A. Rényi, *Probability Theory* (North Holland, Amsterdam, 1970).
31. C. Tsallis, *J. Stat. Phys.* **52**, 479 (1988).
32. C. Holzhey, F. Larsen and F. Wilczek, *Nucl. Phys. B* **424**, 443 (1994).
33. P. Calabrese and J. Cardy, *J. Stat. Mech.* P06002 (2004).
34. F. Franchini, A. R. Its and V. E. Korepin, *J. Phys. A: Math. Theor.* **41**, 025302 (2008).

Chapter 10

SOLITONS EXPERIENCE FOR BLACK HOLE PRODUCTION IN ULTRARELATIVISTIC PARTICLE COLLISIONS

I. YA. AREF'EVA

Steklov Mathematical Institute, Russian Academy of Sciences,

Gubkina Str. 8, 119991, Moscow, Russia

We discuss the analogy between soliton scattering in quantum field theory and black hole/wormholes (BH/WH) production in ultrarelativistic particle collisions in gravity. It is a common wisdom of the current paradigm suggests that BH/WH formation in particles collisions will happen when a center-mass energy of colliding particles is sufficiently above the Planck scale (the transplanckian region) and the BH/WH production can be estimated by the classical geometrical cross section. We compare the background of this paradigm with the functional integral method to scattering amplitudes and, in particular, we stress the analogy of the BH production in collision of ultrarelativistic particle and appearance of breathers poles in the scattering amplitudes in the Sin–Gordon model.

Keywords: Solitons; blackhole production; high energy scattering; extra dimensions.

1. Introduction

Black holes (BHs) formation in collisions of transplanckian particles is one of outstanding problems in theoretical physics. Our aim in this contribution is to discuss a similarity of this problem with the problem of quantum solitons and their scattering in QFT.[1–3]

We apply the scattering amplitudes functional integral method[4,6] to the problem of BHs formation in collisions of transplanckian particles. In particular, we stress the analogy of the BH production in collision of ultrarelativistic particle and appearance of breathers poles in the scattering amplitudes in the Sin–Gordon model.[7]

Study of transplanckian collisions in gravity has a long history. In 80's–90's the problem has been discussed mainly in superstring theory frameworks[8–15] and was considered as an academical one, since the four dimensional Planck scale E_{Pl} is $\approx 10^{19}$ GeV, and energies satisfying $\sqrt{s} > E_{Pl}$ wholly out of reach of terrestrial experiments.

The situation has been changed after the proposal of TeV gravity scenario.[16,17] The D-dimensional Planck energy $E_{Pl,D}$ plays the fundamental role in TeV gravity, it has the electroweak scale of $\sim$ TeV, as this would solve the hierarchy problem. TeV gravity is strong enough to play a role in elementary particle collisions at accessible energies.

The TeV gravity assumes the brane world scenario[18] that means that all light particles (except gravity) are confined to a brane with the 4-dimensional world sheet embedded in the D-dimensional bulk. The collider signatures of such braneworld scenarios would be energy nonconservation due to produced gravitons escaping into the bulk, signatures of new Kaluza–Klein particles as well as signatures of BH[19–23] and more complicated objects such as wormholes (WH) formations.[24–26]

According the common current opinion the process of BH formation in transplanckian collision of particles may be adequately described using classical general relativity. Calculations based on classical general relativity support[27,28] the simple geometrical cross section of BH production in particles collisions, which is proportional to the area of the disk

$$\sigma = f\pi R_S^2(E)\,,\tag{1}$$

where R_S is the Schwarzschild radius of the BH formed in the particles scattering process and it is defined by the center-of-mass collision energy $E = \sqrt{s}$, and f is a formation factor of order unity.

2. Transition Amplitudes and Cross Section of the BH/WH Production

Quantum mechanical formula for the transition amplitude between two states $|A\rangle$ and $|B\rangle$ is given by

$$\langle A|B\rangle = \int \Psi_A^*(X_A,t)\mathcal{K}(X_A,t;X_B,t')\Psi_B(X_B,t')dX_AdX_B\,,\tag{2}$$

where X are generalized coordinates, specifying the system, $\Psi_A(X,t)$, is a wavefunction of the state A including its asymptotical dynamics. The transition amplitude $\mathcal{K}(X_A,t;X_B,t')$ in the generalized coordinate representation is given by the Feynman integral.

According the functional integral method[4,6] the scattering amplitude can be also given by the functional integral

$$\int \exp\left\{\frac{i}{\hbar}S[\phi]\right\}d\phi\,,\tag{3}$$

where we integrate over the field histories with prescribed asymptotic behavior. When solitons (localized classical solutions) are present they have to be included in the list of these asymptotics.[1–3]

In our case we deal not only with particles but also with gravity. In particular, we discuss the process where the final state $|B\rangle$ is the state corresponding to the BH. To this purpose we use a modification[15] of the standard formula (3):

- For simplicity we work in $1 + 3$ formalism where spacetime is presented as a set of slices (more general formulation is described in Ref. 15). At the initial time t we deal with a slice Σ and at the final time t' with a slice Σ'.
- Generalized coordinates include a metric g and matter fields ϕ.
- The state at on a initial time is specified by a three-metric h_{ij} and field ϕ and final state by a three-metric h'_{ij} and ϕ'.
- The transition amplitude in this generalized coordinate representation is given by Feynman integral[15]

$$\mathcal{K}(h, \phi, t; h', \phi', t') = \int e^{\frac{i}{\hbar} S[g, \phi]} \prod_{\substack{\phi|_{\tau=t}=\phi, g|_{\tau=t}=h \\ \phi|_{\tau=t'}=\phi', g|_{\tau=t'}=h'}} \mathcal{D}\phi(\tau) \mathcal{D}g(\tau), \quad (4)$$

where the integral is over all four-geometries and field configurations which match given values on two spacelike surfaces, Σ and Σ' and matter on them, $S[g, \phi]$ is the action. The integral in (4) includes also summation over different topologies.
- The transition amplitude given by the functional integral includes gauge fixing and Faddeev–Popov ghosts (all these are omitted in (4) for simplicity).
- We are interested in the process of a BH creation in particles collisions. Therefore,

  — we specify the initial configuration h and ϕ on Σ without BH, i.e., causal geodesics starting from Σ rich the future null infinity.[a];
  — we specify the final configuration h' and ϕ' on Σ' as describing BH, i.e., Σ' contains a region from which the light does not rich the future null infinity.[b]

The explanation of notions used in above footnotes is given in Ref. 15, see also Appendix. For more details see Refs. 29 and 30.

In Fig. 1 a slice with two colliding particles at $\tau = t$, and $\tau = t'$ with the BH area are presented. To describe such a process in the framework of a general approach (4) we have to find a classical solution of the Einstein equations with the matter, our moving particles, that corresponds to this picture, Fig. 1, and then study quantum fluctuations. We do not have analytical solutions describing this process.

Finding of such solutions is a very difficult problem. It is solved only at low-dimensional case.[31] In 4-dimensional case this problem has been solved numerically only recently by Choptiuk and Pretorius.[32] The solution, as it has been mentioned in Introduction, assumes a construction of a model for gravitational particles. We present this construction in the next subsection.

---

[a]More precise, this condition means that Σ is a partial Cauchy surface with asymptotically simple past in a strongly asymptotically predictable space–time.
[b]This means that Σ' is a partial Cauchy surface containing BH(s), i.e., $\Sigma' - J^-(\mathcal{T}^+)$ is nonempty.

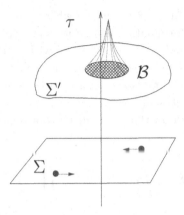

Fig. 1.   A slice Σ at $\tau = t$ is an initial slice with particles and a slice Σ' at $\tau = t'$ is a slice with a BH $\mathcal{B}$. Null geodesics started from the shaded region do not reach null infinity.

3. Ultrarelativistic Particles on the Brane

3.1. *D-dimensional Planckian energy*

In TeV gravity scenario we assume that all particles and fields (except gravity) are confined to a brane with 4-dimensional world sheet embedded in the D-dimensional bulk. Matter fields leave on the brane and do not feel extra dimensions, only gravity feels $n = D - 4$ extra dimensions, see Fig. 2.

According the common current opinion the process of BH formation in transplanckian collision of particle, i.e., in regions where,

$$\sqrt{s} \gg E_{\mathrm{Pl}}, \tag{5}$$

may be adequately described using classical general relativity. We also believe that the same concerns the WH production. This is because in the transplanckian region (5) the de Broglie wavelength of a particle

$$l_{\mathcal{B}} = \frac{\hbar c}{E} \tag{6}$$

is less than the Schwarzschild radius corresponding to this particle,

$$l_{\mathcal{B}} \ll R_{S,D}, \tag{7}$$

here $R_{S,D}$ is the D-dimensional Schwarzschild radius in TeV gravity. In phenomenologically reasonable models with $n \geq 2$ the Schwarzschild radius corresponding to

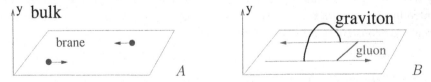

Fig. 2.   (A) Colliding particles on the brane. (B) D-dimensional graviton and 4-dimensional gluon exchanges.

colliding particles with energies ≈ 1 TeV is $R_{S,D} \gtrsim 10^{-16}$ cm. In the usual 4-dimensional gravity the Schwarzschild radius corresponding to the same particles is of order $R_{S,4} \sim 10^{-49}$ cm, that is a negligible quantity comparing with the de Broglie wavelength of particles with energy about few TeV.

Although these type of processes are classical it is instructive to have a full picture starting from a general quantum field theory setup and pass explicitly to the classical description of processes in question. This point of view is useful to deal with effects on the boundary of the classical applicability. By this reason we start in the next subsection from this general setup,[15] and in Sec. 8 we present the brane extension of this approach.[20]

3.2. *D-dimensional gravitational model of relativistic particles*

To start a classical description of BH production in collision of elementary particles we need a gravitational model of relativistic particles. At large distances the gravitational field of particle is the usual Newtonian field. The simplest way to realize this is just to take the exterior of the Schwarzschild metric, i.e., in D-dimensional case away a particle we expect to have:

$$ds^2 = \left(1 - \left(\frac{R_{S,D}}{R}\right)^{D-3}\right) dt^2 + \left(1 - \left(\frac{R_{S,D}}{R}\right)^{D-3}\right)^{-1} dR^2 + R^2 d\Omega_{D-2}^2, \quad (8)$$

where $R_{S,D}$ is the Schwarzschild radius

$$R_{S,D}^{D-3}(m) = \frac{16\pi G_D m}{(D-2)\Omega_{D-2}} = \frac{2m}{(D-2)\Omega_{D-2}M_{\text{Pl},d}^{D-2}}, \quad (9)$$

here G_D is D-dimensional Newton gravitational constant, c the speed of light (in almost all formula we take $c = 1$) and Ω_{D-2} is the geometrical factor,

$$\Omega_{D-2} = \frac{2\pi^{(D-1)/2}}{\Gamma[(D-1)/2]}, \quad (10)$$

Γ is Euler's Gamma function. Here we present D-dimensional formula, in particular for $D = 4$, $R_{S,4}(m) = 2G_4 m$. We also use the expression of the Schwarzschild radius in term of the Planck mass, $R_{S,4}(m) = m/4\pi M_4^2 = m/\bar{M}_4^2$.

The interior of the Schwarzschild metric is supposed to fill with some matter. The simplest possibility is just to take a Tolman–Florides interior incompressible perfect fluid solution.[33,34] As another model of relativistic particles one can consider a static spherical symmetric solitonic solution of gravity-matter equations of motion, the so-called boson stars (authors of Refs. 35 and 36 deal with 4-dimensional space–time, but it not a big deal to get D-dimensional extensions).

In the case of brane scenario few comments are in order. In the simplest brane models we deal with matter only on the brane and we do not have matter out of the brane to fill the interior of the D-dimensional Schwarzschild solution. However,

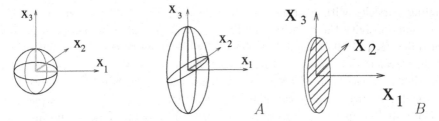

Fig. 3.   (A) Flattening of the Schwarzschild sphere in the boosted coordinates. (B) Schematic picture for the shock wave as a flat disk.

in the string scenario[c] there are closed string excitations which are supposed to be available in the bulk. One can assume that the matter in the bulk is a dilaton scalar field and deal with string inspired D-dimensional generalization boson stars

$$ds^2 = \left(1 - \left(\frac{\mathcal{A}(R)}{R}\right)^{D-3}\right) dt^2 + \left(1 - \left(\frac{\mathcal{B}(R)}{R}\right)^{D-3}\right)^{-1} dR^2 + R^2 d\Omega^2 \,, \qquad (11)$$

where

$$\mathcal{A}(R) = A + A_1/R + \cdots . \qquad (12)$$

4. Shock Wave as a Model of Ultra Relativistic Moving Particle

To consider ultra relativistic moving particle we have to make a boost of metric (11) with the large Lorentz boost factor $\gamma = 1/\sqrt{1 - v^2/c^2}$. The Schwarzschild sphere under this boost flattens up to an ellipsoid, see Fig. 3.

One can consider an approximation when γ is taken infinitely large and $E = \gamma A$ is fixed. The result metric is the Aichelburg–Sexl (AS) metric,[37,38] a gravitational shock wave, where the nontrivial geometry is confined to a D-2-dimensional plane traveling at the speed of light, with Minkowski spacetime on either side,

$$ds^2 = -2dU\,dV + dX_i^2 + F(X)\delta(U)dU^2 \,, \quad i = 2, 3, \dots, D-1 \,, \qquad (13)$$

where $V = (X^0 + X^1)/\sqrt{2}$, $U = (X^0 - X^1)/\sqrt{2}$. The form of the profile of the shock wave F depends on the behavior of $\mathcal{A}(r)$.

In particular, in the infinite boost limit where we also take $m \to 0$ and hold p fixed, the metric (8) reduces to an exact shock wave metric (13) with the shape function F being the Green function of the D-2-dimensional Laplace equation

$$\Delta_{R^{D-2}} F = -\frac{2p\sqrt{2}}{M_{\text{Pl}}^{D-2}} \delta^{(D-2)}(X) \,, \qquad (14)$$

here $\delta^{(D-2)}(X) = \prod_{i=2}^{D-1} \delta(X^i)$,

$$F(X) = \frac{p2\sqrt{2}}{(D-4)\Omega M_{\text{Pl},D}^{D-2}} \frac{1}{\rho^{D-4}} \,, \qquad (15)$$

---

[c]Open string excitations are located on brane, closed string excitations propagate on the bulk.

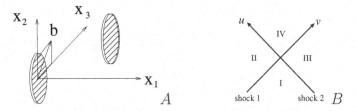

Fig. 4. (A) Ultra relativistic colliding particles in D-1-dimensional space; b is the D-2-dimensional impact vector. (B) Ultra relativistic colliding particles in U, V-plane.

where $\rho^2 = (X^2)^2 + \cdots (X^{D-1})^2$. For $D = 4$ the shape is:

$$F(X) = -\frac{p\sqrt{2}}{\pi M_4^2} \ln \frac{\rho}{\varepsilon}. \qquad (16)$$

Note, that the metric (13) is obtained in the infinite boost limit when the source has zero rest mass. For fast particles of nonzero rest mass, the shock wave approximation breaks down far away from the moving particle, more precisely at transverse distances from the source which are of the order of:

$$\ell \sim r_h(m)/\sqrt{1 - v^2}. \qquad (17)$$

At these distances the field lines will spread out of the null transverse surface orthogonal to the direction of motion. But for $b \ll \ell$ one can use the shock wave field to extract the information about the BH formation to the leading order in m/p. These shock wave are presented in Fig. 4 as sphere flattened up to the disk. Two such shock waves, moving in opposite directions, see Fig. 4(B) give the pre-collision geometry of the spacetime. Though the geometry is not known to the future of the collision, since the shock wave solutions inevitably break down when the fields of different particles cross, at the moment of collision a trapped surface can be found.[27,28]

According to Refs. 27 and 28, the trapped surfaces do form when $b \lesssim R_{S,D}$, and have the area of the order $\sim R_{S,D}^2$, where $R_{S,D}$ is the horizon radius given by (19).

Infinitely thin shock is an idealization. In reality the shocks will have a finite width w since γ is large but not infinite. The corresponding shocks have width $w_{\text{class}} \sim r/\gamma$, depending on the transverse distance r. Infinitely thin idealization leads to an appearance in the intersection of the planes of the two shock waves a divergent curvature invariant.[39] In Ref. 40 has shown that this problem is an artifact of the unphysical classical point-particle limit and for a particle described by a quantum wavepacket, or for a continuous matter distribution, trapped surfaces indeed form in a controled regime.

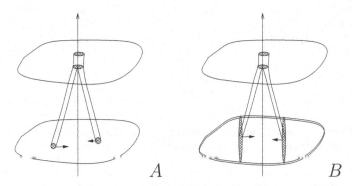

Fig. 5.   (A) Colliding two stars; the initial space-time is asymptotically flat. (B) Colliding shock waves as models of ultra relativistic particles; the initial space time is not asymptotically flat.

5. D-Dimensional Thorne Hoop Conjecture and Geometrical Cross Section

We expect to get the BH formation due to nonlinear interaction of gravitational fields produced by particles. The BH formation in classical general relativity is controlled by the Thorne hoop conjecture.[41] According to the D-dimensional version of this conjecture if a total amount of matter mass M is compressed into a spherical region of radius R, a BH will form if R is less than the corresponding Schwarzschild radius

$$R < R_{S,D}(M),\qquad(18)$$

here $R_{S,D}(M)$ is given by (9).

In the case of ultra relativistic particle collisions the main argument for BH formation is based on a modification of Thorne's hoop conjecture. According to this modified conjecture if a total amount of energy E is compressed into a spherical region of radius R, a BH will form if R is less than the corresponding Schwarzschild radius

$$R < R_{S,D}(E) \equiv \left(\Omega_n \frac{G_D E}{c^4}\right)^{\frac{1}{n+1}}.\qquad(19)$$

Note that in this modified conjecture the horizon radius $R_{S,D}$ is set by the center-of-mass collision energy $E = \sqrt{s}$.

Few remarks are in order concerning this formulation. Literally speaking, as it is formulated above, it is not applicable in all situations. But this conjecture does apply for two colliding particles. There are several calculations and arguments supporting this conjecture:

- One set of arguments is related with examining trapped surface formation in collisions of ultra relativistic particles.[23,27] Note that commonly used evidence for BH formation in collision of particles comes from the study of the collision of two AS shock wave. This argument assumes that there is a solution interpolating

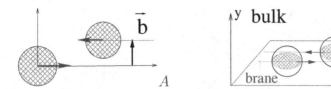

Fig. 6.   (A) Colliding particles in D-1-dimensional space: b is D-2-dimensional impact vector and $\sigma \approx \mathcal{D}R_{S,D}^{D-2}$. (B) Colliding particles on the 3-brane: 2-dimensional impact vector b and $\sigma \approx \pi R_{S,D}^2$. A shaded region indicates the projection of D-1-dimensional Schwarzschild sphere onto the 3-brane.

between two shock waves and BH, Fig. 5(B). However with this argument there is a problem that a space time with a shock wave is not asymptotically flat, that assumed in our scheme.[d]

- The same problem is also with colliding plane wave.[15] An advantage to deal with plane waves is that in this case one can construct explicitly the metric in the interacting region.
- There is a nontrivial possibility to reduce the proof Thorne's hoop conjecture for ultra colliding relativistic particles to Thorne's hoop conjecture for slow moving relativistic particles.[63]
- There are resent numerical calculations supporting (19).[32] In Ref. 32 as a model of particles the boson star is taken.[35] Choptiuk and Pretorius have got a remarkable result that BH do form at high velocities in boson star collisions and they found also that this happens already at a γ-factor of roughly one-third predicted by the hoop conjecture.

On the modified Thorne's hoop conjecture for ultra colliding relativistic particles the so-called geometrical cross section of BH production is based. It estimates the BH production cross section by the horizon area of a BH whose horizon radius $R_{S,D}$ is set by the center-of-mass collision energy $E = \sqrt{s}$, Eq. (19). This estimation assumes that when the impact parameter b is smaller than $R_{S,D}$ then the probability of formation of a BH is close to one,

$$\sigma_{BH,D} \approx \mathcal{D}_{D-2} R_{S,D}^{D-2}(E) , \qquad (20)$$

$\mathcal{D}_{D-2}$ is the volume of a plane cross section of the D-2-dimensional unit sphere, see Fig. 6(A) where b is D-2-dimensional vector, i.e., the area of of D-2-dimensional disk,

$$\mathcal{D}_D = \frac{\pi^{D/2}}{\Gamma\left(1 + \frac{D}{2}\right)} . \qquad (21)$$

---

[d]The AS metric also has a naked singularity at the origin. This is considered as an artifact of having used a BH metric as the starting point, and assumed to be removed by taking a suitable mass distribution.

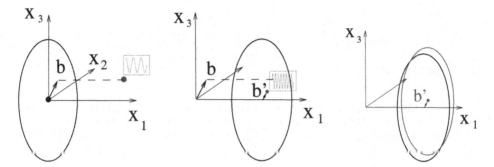

Fig. 7.   (A) Ultra relativistic particle (shock wave) and a rest quantum particle, **b** is an impact vector. (B) Quantum particle after a collision with the ultra relativistic particle, its impact vector **b′** just after collision decreases, $|\mathbf{b}'| \ll |\mathbf{b}|$ and its frequency increases. (C) After collision particle which was in rest after collision move with an ultra relativistic velocity and looks as a shock wave.

In the 4-dimensional case this estimation gives

$$\sigma_{BH,4} \approx \pi R_{S,4}^2(E)\,. \tag{22}$$

For the 3-brane embedding in the D-dimensional space–time we have

$$\sigma_{BH,\text{3-brane}} \approx \pi R_{S,D}^2(E) \tag{23}$$

since our particles are restricted on the 3-dimensional brane and the impact vector **b** is two dimensional vector, see Fig. 6(B).

6. Looking from the Rest Frame of One of the Incident Particles

It is instructive to note that the similar analyze can be done in the rest frame of one of the incident particles.[38] This particle has large de Broglie wavelength and has to be treated as a quantum particle. The gravitational field of the other, which is rapidly moving, looks like a gravitational shock wave, see Fig. 7.

Dynamics of the quantum particle can be described by a solution of the quantum Klein–Gordon equation in the shock wave background. This problem has been solved by 't Hooft.[53,54] Dynamics of the particle is given by the eikonal approximation and is defined by the geodesics behavior near the shock wave. The approximation is valued for a large impact parameter. The shock wave focuses the geodesics down to a small impact parameter. Just in this region we expect the BH formation (see next section) and in this region the eikonal approximation is not nonapplicable. This gives an explanation why a straightforward eikonal approximation does not describe the BH production. But it is instructive to see what the eikonal approximation can give and this is a subject of the next subsection.

The picture presented in Fig. 7 is idealization. More precise approach would be started from one moving particle with γ rather large, but $\gamma \neq \infty$ and other particle in the rest. It should exist a classical solution that interpolates between this initial

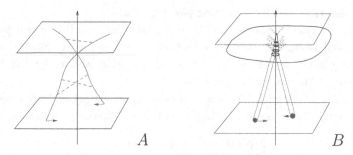

Fig. 8. (A) Ultra relativistic colliding particles with a large impact parameter. Blue lines represent the graviton exchange. (B) Colliding particles with a small impact parameter and mass/energy enough to produce BH. Red dot lines represent BH evaporation.

configuration and a configuration in the later time that represents two stars which are rather closed and move slowly with respect to each other. One can expect to estimate quantum fluctuations to such classical configuration.

7. BH Formation and the Eikonal Approximation

For a large impact parameter (see Figs. 9 and 10) in the transplanckian region one can use the eikonal approximation and for the graviton exchange diagrams we get,[45,55,56]

$$\mathcal{A}_{\text{eik}}(\mathbf{q}) = \mathcal{A}_{\text{Born}} + \mathcal{A}_{1-\text{loop}} + \cdots = -8Ep \int d^2\mathbf{b}\, e^{-i\mathbf{q}\cdot\mathbf{b}}(e^{i\chi} - 1)\,, \tag{24}$$

with the eikonal phase χ given by the Fourier transform of the Born amplitude in the transverse plane. The 4-dimensional Born amplitude for the graviton exchange is given by:

$$\mathcal{A}_{\text{Born}}(\mathbf{q}) = \frac{2\pi G\gamma(s)}{Ep}\frac{1}{q_\perp^2 + \mu^2} \tag{25}$$

here μ is IR graviton mass regularization. The corresponding eikonal phase,[55] is:

$$\chi = \frac{2\pi G\gamma(s)}{Ep} \int \frac{d^2 q_\perp}{(2\pi)^2} e^{iq_\perp \cdot x_\perp}\frac{1}{q_\perp^2 + \mu^2} = \frac{2\pi G\gamma(s)}{Ep}K_0(\mu b)\,, \tag{26}$$

where $\gamma(s) = 1/2((s - 2m^2)^2 - 2m^4)$, K_0 is the modified Bessel function.

For $b\mu \ll 1\, K_0(\mu b) \sim (1/4\pi)\ln(\mu b)$ and we get the eikonal amplitude in term of Mandelstam variables

$$\mathcal{A}_{\text{eik}}(\mathbf{q}) = \frac{16\pi G\gamma(s)}{-t}\frac{\Gamma(1 - i\alpha(s))}{\Gamma(1 + i\alpha(s))}\left(\frac{4\mu^2}{-t}\right)^{-i\alpha(s)}\,, \quad \alpha(s) = \frac{2G\gamma(s)}{\sqrt{s(s - 4m^2)}} \tag{27}$$

The eikonal approximation with a real eikonal phase satisfies the unitarity condition

$$\sigma_{\text{eik}} = \frac{1}{16\pi^2 s^2}\int d^2 q_\perp |\mathcal{A}_{\text{eik}}|^2 = \frac{\text{Im}\mathcal{A}_{\text{eik}}(0)}{s}\,, \tag{28}$$

and cannot describe the BH formation [see Fig. 8(a)].

However, the exact knowledge of $2 \to 2$ scattering amplitude can provide information about resonance states production. Therefore, if we know exact $2 \to 2$ scattering amplitude we expect to get an information about the BH formation [Fig. 8(b)] due to the unitarity condition.

There is an analogy with BH production in higher energy and breather production in the two particles scattering in Sin–Gordon 2-dimensional model,

$$2 \text{ particles } \to \text{ breather}. \tag{29}$$

Indeed, the classical Sin–Gordon 2-dimensional model has so-called breather solutions with masses

$$m_n = \frac{16m}{\gamma} \sin \frac{n\gamma}{16}, \quad n = 1, \dots < \frac{8\pi}{\gamma}. \tag{30}$$

The exact quantum $2 \to 2$ amplitude $\mathcal{A}_{\text{exact}}$ for massive particles in the 2-dimensional Sin–Gordon model[7] has an extra pole at:

$$M^2 = 4m_1^2 - m_1^2 \left(\frac{\gamma}{8}\right)^2 + \cdots \tag{31}$$

that is nothing but the pole corresponding to the first breather. One can see the breather contribution in the unitarity condition for amplitude of massive particles, $\mathcal{A}_{\text{exact}}$.

BH production in the collision of two particles can also seen as a violation of the unitary in the $2 \to 2$ elastic channel. Indeed, let us consider a scattering amplitude in two channels system,

$$\mathcal{A} = \begin{pmatrix} \mathcal{A}_{2p \to 2p} & \mathcal{A}_{2p \to \text{BH}} \\ \mathcal{A}_{\text{BH} \to 2p} & \mathcal{A}_{\text{BH} \to \text{BH}} \end{pmatrix}. \tag{32}$$

$\mathcal{A}_{2p \to 2p}$ is the elastic scattering amplitude and $\mathcal{A}_{2p \to \text{BH}}$ is the inelastic one. The unitary condition means that:

$$2\text{Im}\,\mathcal{A}_{2p \to 2p} = |\mathcal{A}_{2p \to 2p}|^2 + |\mathcal{A}_{2p \to \text{BH}}|^2. \tag{33}$$

So, if we expect $\mathcal{A}_{2p \to \text{BH}} \neq 0$ we have a violation of the the elastic unitarity,

$$2\text{Im}\,\mathcal{A}_{2p \to 2p} \neq |\mathcal{A}_{2p \to 2p}|^2. \tag{34}$$

The simple way to break unitarity is to assume that in the eikonal approximation we deal with a complex eikonal phase. We expect the imaginary eikonal phase at small impact parameter and we write:

$$\mathcal{A}_{\text{eik}}^{(2 \to 2)}(\mathbf{q}) = -2s \int_{|\mathbf{b}|>b_c} d^2\mathbf{b}\, e^{-i\mathbf{q}\cdot\mathbf{b}} (e^{i\chi} - 1)$$

$$- 2s \int_{|\mathbf{b}|<b_c} d^2\mathbf{b}\, e^{-i\mathbf{q}\cdot\mathbf{b}} (e^{-\delta+i\chi} - 1), \tag{35}$$

$b_c \sim R_{S,4}$. We have:

$$\sigma_{\text{el}} = \frac{1}{16\pi^2 s^2} \int \frac{d^2\mathbf{q}}{(2\pi)^2} |\mathcal{A}_{\text{eik}}^{(2\to2)}|$$

$$= 2 \int_{|\mathbf{b}|>b_c} d^2\mathbf{b}\,[1 - \cos\chi] + \int_{|\mathbf{b}|<b_c} d^2\mathbf{b}\,[1 + e^{-2\delta} - 2e^{-\delta}\cos\chi]. \tag{36}$$

In accordance with the optical theorem,

$$\sigma_{\text{total}} = \frac{1}{s}\text{Im}\,\mathcal{A}_{\text{eik}}^{(2\to2)}(0)$$

$$= 2\int_{|\mathbf{b}|>b_c} d^2\mathbf{b}\,[1 - \cos\chi] + 2\int_{|\mathbf{b}|<b_c} d^2\mathbf{b}\,[1 - e^{-\delta}\cos\chi] \tag{37}$$

and one can interpret the difference between (36) and (37) as a cross section of the BH production[45,50,59]

$$\sigma_{\text{BH}} = \sigma_{\text{total}} - \sigma_{\text{el}} = \int_{|\mathbf{b}|<b_c} d^2\mathbf{b}[1 - e^{-2\delta}] \tag{38}$$

To summarize the above discussion we can say that to describe the BH creation we would need to use the full classical solution describing the process, which however is difficult to handle. From other site, the full $2 \to 2$ particle amplitude would provide information about the BH production, but we are faraway from getting it. The elastic small-angle scattering amplitude given by eikonalized single-graviton exchange,[11,12,53–55,57] valued for large impact parameters $b \gg R_{S,4}$, cannot describe the BH formation. Computing the corrections in $b/R_{S,4}$ to the elastic scattering, one hopes to learn about the strong inelastic dynamics at $b \sim R_{S,4}$.[11,12,48,49,58]

8. Transition Amplitudes and Cross Section for Higher Dimensional Gravity and Matter Living on the Brane

To consider the question of BH creation in high energy scattering in physical setting for low Planck scale we have to deal with two particles confined to the 3-brane which scatter due to the D-dimensional gravitational field, $D = 4 + n$, n is the number of large extra dimensions (see Figs. 9 and 10). For this purpose we have to make few modifications of formula (4) and take into account that particles interact with D-dimensional gravity and with matter leaving on the brane.

- For simplicity we work in $1 + n$ formalism. At the initial time t we deal with a slice Σ and at the final time t' with a slice Σ'. The slice Σ crosses the brane worldsheet over 3-dimensional slice Ξ and the slice Σ crosses the brane worldsheet over 3-dimensional slice Ξ'.
- Generalized coordinate include D-dimensional metric g_{MN} and matter fields ϕ leaving on the brane B.

- The state at a initial time is specified by a $3 + n$-metric h_{IJ} on the slice Σ and fields ϕ on the slice Ξ and final state by a $3 + n$-metric h'_{IJ} on the slice Σ' and ϕ' on the slice Ξ'.

- The transition amplitude in this generalized coordinate representation is given by Feynman integral, which is an extension of the formula from Ref. 15 to the brane world

$$\mathcal{K}(h, \phi, t; h', \phi', t') = \int e^{\frac{i}{\hbar}S[g,\phi]} \prod_{\substack{\phi|_{\tau=t}=\phi,\, g|_{\tau=t}=h \\ \phi|_{\tau'=\phi'},\, g|_{\tau'=h'}}} \mathcal{D}\phi(\tau, \mathbf{x})\mathcal{D}g(\tau, \mathbf{X})$$

where the integral is over all $4 + n$-geometries which match given values on two spacelike surfaces, Σ and Σ' and field configurations which match given values on two 3-dimensional spacelike surfaces, Ξ and Ξ'.

- We specify the initial configuration h so that it corresponds to the Minkowski brane embedding in the bulk and matter fields ϕ on Ξ;

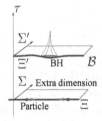

A slice Σ' at $\tau = t'$ is a slice with a BH $\mathcal{B}$.
Null geodesics started from the shaded region on Ξ' do not reach null infinity.

A slice Σ at $\tau = t$ is an initial slice with extra dimensions and particles living on the brane Ξ (blue thick line).

- We specify the final configuration h' on Σ' and ϕ' on Ξ' as describing black hole. $\mathcal{B}$ is a in D-dimensional BH.

It is not simple to a find solution with a D-dimensional BH and a brane. The reason is that the usual D-dimensional BH, say the Meyer–Perry BH, solves the vacuum D-dimensional Einstein equation. But in the case of the presence of the brane the energy momentum tensor has an extra term providing the localization of the matter on the brane,[52,60–62] $T_{MN} \sim \delta(y)t_{MN}$.[63]

As a conclusion, let us note that recently the considerations of the BH production in collisions of shock waves in AdS space–time have got applications in the observable variable that is the multiplicity of particles production in collisions of quark gluon plasma.[64–71]

Acknowledgments

It is my pleasure to present this paper to the proceedings dedicated to Volodya Korepin's sixtieth birthday.

This work is supported in part by RFBR Grant 11-01-00894-a.

Appendix A. BH as an Initial/Final Data

Let $(\mathcal{M}, g)$ is the space–time with a metric, $\mathcal{M}$ is a manifold.

The BH are conventionally defined in asymptotically flat space–times by the existence of an event horizon H.

The horizon H is the boundary of the causal past of future null infinity, i.e., it is the boundary of the set of events in space–time from which one can escape to infinity in the future direction.

To be more precise we need few definitions.[29,30]

All notions work for BH production in the particle collision as well as for gravitational collapse.

A.1. *(Weakly) asymptotically simple space–time*

A (oriented in time and space) space–time $(\mathcal{M}, g_{\mu\nu})$ is *asymptotically simple* (Ref. 29, p. 246) if there exists a smooth manifold $\tilde{\mathcal{M}}$ with metric $\tilde{g}_{\mu\nu}$, boundary $\mathcal{I}$, and a scalar function Ω regular everywhere on $\tilde{\mathcal{M}}$ such that:

- $\tilde{\mathcal{M}} - \mathcal{I}$ is conformal to $\mathcal{M}$ with $\tilde{g}_{\mu\nu} = \Omega^2 g_{\mu\nu}$,
- $\Omega > 0$ in $\tilde{\mathcal{M}} - \mathcal{I}$ and $\Omega = 0$ on $\mathcal{I}$ with $\nabla_\mu \Omega \neq 0$ on $\mathcal{I}$,
- Every null geodesic on $\tilde{\mathcal{M}}$ contains, if maximally extended, two end points on $\mathcal{I}$.

$\mathcal{I}$ consists of two disjoint pieces $\mathcal{I}^+$ (future null infinity) and $\mathcal{I}^-$ (past null infinity) each topologically $R \times S^2$,

$$\mathcal{I} = \mathcal{I}^+ \cup \mathcal{I}^- . \tag{39}$$

One can symbolically write

$$\tilde{\mathcal{M}} = \mathcal{M} \cup \partial \mathcal{M} , \tag{40}$$

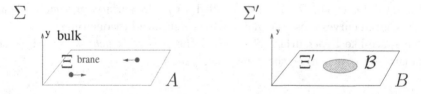

Fig. 9.   Slices with brane at different times. (A) Initial slice Σ with brane Ξ and particles on the brane and without BH; (B) Finite slice Σ' with a BH on the brane.

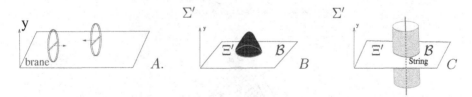

Fig. 10.   (A) Ultra relativistic colliding particles on the 3-brane; a blue shaded region corresponds to a cross section of the D-2-dimensional disk by the 3-brane; (B, C). Slices with brane at final time. (B) BH with a source localized on a point at the brane Ξ'; (C) Black string with a source localized on a line along extra dimensions.

where $\partial \mathcal{M} = \mathcal{I}$.

If M satisfies the Einstein vacuum equations near $\mathcal{I}$ then $\mathcal{I}$ is null.

A space–time $\mathcal{M}$ is *weakly asymptotically simple* if there exists an asymptotically simple M_0 with corresponding $\tilde{\mathcal{M}}_0$ such that for some open subset K of $\tilde{\mathcal{M}}_0$ including $\mathcal{I}$, the region $\mathcal{M}_0 \cup K$ is isometric to an open subset of $\mathcal{M}$. This allows $\mathcal{M}$ to have more infinities than just $\mathcal{I}$.

A.2. *Asymptotically flat space–time*

A space–time is *asymptotically flat* if it is *weakly asymptotically simple and empty*, that is, near future and past null infinities it has a conformal structure like that of Minkowski space–time.

A.3. *Global (partial) Cauchy surface*

Let $\mathcal{S}$ be a space-like hypersurface. The future (past) domain of dependence of $\mathcal{S}$, denoted $D^+(\mathcal{S})(D^-(\mathcal{S}))$, is defined by[29]:

$$D^{\pm}(\mathcal{S})$$
$$= \{p \in \mathcal{M} | \text{Every past (future) inextensible causal curve through } p \text{ intersect } \mathcal{S}\} \tag{41}$$

The full domain of dependence of $\mathcal{S}$ is defined as:

$$D(\mathcal{S}) = D^+(\mathcal{S}) \cup D^-(\mathcal{S}). \tag{42}$$

The set $\mathcal{S}$ for which $D(\mathcal{S}) = \mathcal{M}$ is called a Cauchy surface, or a *global Cauchy surface*. Causal curves cross the global Cauchy surface just one time.

The space-like hypersurface $\mathcal{S}$ is called the *partial Cauchy surface* if nonone causal curve crosses it more then one time,[29] see Fig. 11.

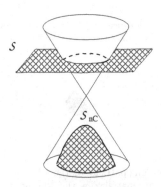

Fig. 11.   (A) The Minkowski space–time M^4 with the Cauchy surface $\mathcal{S}$ and a surface $\mathcal{S}_{nC}$ which is not the Cauchy surface, Ref. 29, Fig. 13

A.4. *Causal future $J^+(p)$ and causal past $J^-(p)$ of the point*

A causal future $J^+(p)$ (past $J^-(p)$) of the point p is defined as:

$$J^\pm(p) = \{q \in \mathcal{M}, \text{ such that there is future (past) oriented}$$
$$\text{causal curve } \gamma(\tau), \text{ so that } \gamma(0) = p, \gamma(1) = q\}. \tag{43}$$

A causal future $J^+(\Sigma)$ of the surface Σ is defined as:

$$J^-(\Sigma) = \{q \in \mathcal{M}, \text{ if there is future oriented}$$
$$\text{causal curve } \gamma(\tau), \text{ so that } \gamma(0) = p \in \Sigma, \ \gamma(1) = q\}. \tag{44}$$

In particular,

$$J^-(\mathcal{I}^+) = \{q \in \mathcal{M}, \text{ if there is future oriented}$$
$$\text{causal curve } \gamma(\tau), \text{ so that } \gamma(0) = p \in \mathcal{I}^+, \gamma(1) = q\}. \tag{45}$$

A.5. *BHs in asymptotically flat space–time*

The BHs are conventionally defined in asymptotically flat space–times by the existence of an event horizon H. The horizon H is the boundary $\dot{J}^-(\mathcal{I}^+)$ of the causal past $J^-(\mathcal{I}^+)$ of future null infinity $\mathcal{I}^+$, i.e., it is the boundary of the set of events in space–time from which one can escape to infinity in the future direction.

The *BH region B* is $B = M - J^-(\mathcal{I}^+)$ and *the event horizon is* $H = \dot{J}^-(\mathcal{T}^+)$.

A.6. *Future (strongly) asymptotically predictable space–time*

A space–time is *future asymptotically predictable* if there is a surface S in space–time that serves as a Cauchy surface for a region extending to future null infinity.[e]

This means that there are no "naked singularities" (a singularity that can be seen from infinity) to the future of the surface S.

Let Σ be a partial Cauchy surface in a weakly asymptotically simple and empty space–time (M, g). The space–time (M, g) is (future) *asymptotically predictable from* Σ if $\mathcal{I}^+$ is contained in the closure of $D^+(\Sigma)$ in $\tilde{M}_0$.

If, also, $J^+(\Sigma) \cap \bar{J}^-(\mathcal{I}^+, \bar{M})$ is contained in $D^+(\Sigma)$ then the space–time (M, g) is called *strongly asymptotically predictable* from Σ. In such a space there exist a family $\Sigma(\tau)$, $0 < \tau < \infty$, of spacelike surfaces homeomorphic to Σ which cover $D^+(\Sigma) - \Sigma$ and intersects $\mathcal{I}^+$. One could regard them as surfaces of constant time.

A.7. *BH on a surface*

A *BH on the surface* $\Sigma(\tau)$ is a connected component of the set

$$B(\tau) = \Sigma(\tau) - J^-(\mathcal{I}^+, \bar{M}).$$

[e]This notion gives a formulation of Penrose's cosmic censorship conjecture.

One is interested primarily in BHs which form from an initially nonsingular state. Such a state can be described by using the partial Cauchy surface Σ which has an *asymptotically simple past*, i.e., the causal past $J^-(\Sigma)$ is isometric to the region $J^-(\mathcal{I})$ of some asymptotically simple and empty space–time with a Cauchy surface $\mathcal{I}$. Then Σ has the topology R^3.

A.8. *Boundary conditions in the path integral representation*

One is interested primarily in BHs which form from an initially nonsingular state. Such a state can be described by using the partial Cauchy surface Σ which has an *asymptotically simple past*, i.e., the causal past $J^-(\Sigma)$ is isometric to the region $J^-(\mathcal{I})$ of some asymptotically simple and empty space–time with a Cauchy surface $\mathcal{I}$. Then Σ has the topology R^3.

In the path integral representation considered (2) we suppose that we deal with a set of space–times $\{(M, g_{\mu\nu})\}$ which are weakly asymptotically simple and empty and strongly asymptotically predictable. We take into account only such $(M, g_{\mu\nu})$ that contains Σ' and Σ'' so that:

- Σ' is a partial Cauchy surface with asymptotically simple past, $\Sigma' \sim R^3$.
- $\Sigma'' = \Sigma(\tau'')$ contains a BH, i.e., $\Sigma'' - J^-(\mathcal{I}^+, \bar{M})$ is nonempty.

In particular, if one has the condition $\Sigma' \cap \bar{J}^-(\mathcal{I})$ is homeomorphic to R^3 (an open set with compact closure) then Σ'' also satisfies this condition.

References

1. V. E. Korepin, P. P. Kulish and L. D. Faddeev, *JETP Lett.* **21**, 138 (1975) [*Pisma Zh. Eksp. Teor. Fiz.* **21**, 302 (1975)].
2. L. D. Faddeev and V. E. Korepin, *Phys. Lett. B* **63**, 435 (1976).
3. L. D. Faddeev and V. E. Korepin, *Phys. Rept.* **42**, 1 (1978).
4. L. D. Faddeev, *Preprint — FADDEEV L D (78,REC.MAY) 45p*, www.slac.stanford.edu/spires/find/hep/www?iru=553646
5. L. D. Faddeev, Introduction to functional methods, in *Proc. of Ecole d'Ete de Physique Theorique — Methods in Field Theory*, Les Houches, France, 28 Jul–6 Sep 1975, pp. 1–40.
6. I. Y. Arefeva, L. D. Faddeev and A. A. Slavnov, *Theor. Math. Phys.* **21**, 1165 (1975) [*Teor. Mat. Fiz.* **21**, 311 (1974)].
7. I. Arefeva and V. Korepin, *Pisma Zh. Eksp. Teor. Fiz.* **20**, 680 (1974).
8. G. 't Hooft, Gravitational collapse and particle physics, in *Proc.: Proton-Antiproton Collider Physics* (Aachen, 1986), pp. 669–688.
9. G. 't Hooft, *Phys. Lett. B* **198**, 61 (1987).
10. G. 't Hooft, *Nucl. Phys. B* **304**, 867 (1988).
11. D. Amati, M. Ciafaloni and G. Veneziano, *Phys. Lett. B* **197**, 81 (1987).
12. D. Amati, M. Ciafaloni and G. Veneziano, *Int. J. Mod. Phys. A* **3**, 1615 (1988).
13. D. J. Gross and P. F. Mende, *Nucl. Phys. B* **303**, 407 (1988).
14. D. Amati, M. Ciafaloni and G. Veneziano, *Nucl. Phys. B* **403**, 707 (1993).
15. I. Y. Aref'eva, K. S. Viswanathan and I. V. Volovich, *Nucl. Phys. B* **452**, 346 (1995), [arXiv:hep-th/9412157].

16. N. Arkani-Hamed, S. Dimopoulos and G. Dvali, *Phys. Lett. B* **429**, 263 (1998), [arXiv:hep-ph/9803315].
17. I. Antoniadis *et al.*, *Phys. Lett. B* **436**, 257 (1998), [arXiv:hep-ph/9804398].
18. V. A. Rubakov and M. E. Shaposhnikov, *Phys. Lett. B* **125**, 136 (1983).
19. T. Banks and W. Fischler, hep-th/9906038.
20. I. Ya. Aref'eva, *Part. Nucl.* **31**, 169 (2000), hep-th/9910269.
21. S. Dimopoulos and G. Landsberg, *Phys. Rev. Lett.* **87**, 161602 (2001), hep-ph/0106295.
22. P. C. Argyres, S. Dimopoulos and J. March-Russell, *Phys. Lett. B* **441**, 96 (1998), [hep-th/9808138].
23. S. B. Giddings and S. Thomas, *Phys. Rev. D* **65**, 056010 (2002), [arXiv:hep-ph/0106219].
24. I. Y. Aref'eva and I. V. Volovich, *Int. J. Geom. Meth. Mod. Phys.* **05**, 641 (2008), [arXiv:0710.2696 [hep-ph]].
25. A. Mironov, A. Morozov and T. N. Tomaras, arXiv:0710.3395.
26. P. Nicolini and E. Spallucci, arXiv:0902.4654.
27. R. Penrose, unpublished, 1974.
28. D. M. Eardley and S. B. Giddings, *Phys. Rev. D* **66**, 044011 (2002), [gr-qc/0201034].
29. S. W. Hawking and G. R. F. Ellis, *The Large Scale Structure of Space–Time* (Cambridge University Press, UK, 1973).
30. R. M. Wald, *General Relativity* (The University of Chicago Press, Chicago, 1984).
31. G. 't Hooft, *Commun. Math. Phys.* **117**, 685 (1988).
32. M. W. Choptuik and F. Pretorius, arXiv:0908.1780[gr-qc].
33. R. C. Tolman, *Phys. Rev.* **55**, 364 (1939).
34. P. S. Florides, *Proc. R. Soc. London Ser. A* **337**, 529 (1974).
35. M. Colpi, S. L. Shapiro and I. Wasserman, *Phys. Rev. Lett.* **57**, 2485 (1986).
36. F. E. Schunck and E. W. Mielke, *Class. Quant. Grav.* **20**, R301 (2003), [arXiv:0801.0307 [astro-ph]].
37. P. C. Aichelburg and R. U. Sexl, *Gen. Rel. Grav.* **2**, 303 (1971).
38. T. Dray and G. 't Hooft, *Nucl. Phys. B* **253**, 173 (1985).
39. V. S. Rychkov, *Phys. Rev. D* **70**, 044003 (2004), hep-ph/0401116.
40. S. B. Giddings and V. S. Rychkov, hep-th/0409131.
41. K. S. Thorne, Nonspherical gravitational collapse — A short review, in *Magic Without Magic*, ed. J. Klauder (W. H. Freeman, San Francisco, 1972).
42. G. 't Hooft, *Phys. Lett. B* **198**, 61 (1987).
43. D. Kabat and M. Ortiz, hep-th/9203082.
44. D. Amati, M. Ciafaloni and G. Veneziano, *Nucl. Phys. B* **403**, 707 (1993).
45. G. F. Giudice, R. Rattazzi and J. D. Wells, *Nucl. Phys. B* **630**, 293 (2002), [arXiv:hep-ph/0112161].
46. H. L. Verlinde and E. P. Verlinde, *Nucl. Phys. B* **371**, 246 (1992), arXiv:hep-th/9110017.
47. M. Ciafaloni and D. Colferai, *JHEP* **0811**, 047 (2008), arXiv:0807.2117 [hep-th].
48. G. Veneziano and J. Wosiek, *JHEP* **0809**, 023 (2008), arXiv:0804.3321 [hep-th].
49. G. Veneziano and J. Wosiek, *JHEP* **0809**, 024 (2008), arXiv:0805.2973[hep-th].
50. S. B. Giddings and R. A. Porto, arXiv:0908.0004.
51. P. Lodone and S. Rychkov, arXiv: 0909.3519.
52. I. Y. Aref'eva *et al.*, *Nucl. Phys. B* **590**, 273 (2000) [arXiv:hep-th/0004114].
53. G. 't Hooft, *Phys. Lett. B* **198**, 61 (1987).
54. G. 't Hooft, *Nucl. Phys. B* **304**, 867 (1988).
55. D. Kabat, M. Ortiz, hep-th/9203082.

56. D. Amati, M. Ciafaloni and G. Veneziano, *Nucl. Phys.* B **403**, 707 (1993).
57. H. L. Verlinde and E. P. Verlinde, *Nucl. Phys.* B **371**, 246 (1992), arXiv:hep-th/9110017.
58. M. Ciafaloni and D. Colferai, *JHEP* **0811**, 047 (2008), arXiv:0807.2117 [hep-th].
59. P. Lodone and S. Rychkov, arXiv:0909.3519.
60. L. Randall and R. Sundrum, *Phys. Rev. Lett.* **83**, 3370 (1999).
61. T. Shiromizu, K. I. Maeda and M. Sasaki, *Phys. Rev.* D **62**, 024012 (2000), gr-qc/9910076.
62. R. Gregory, V. A. Rubakov and S. M. Sibiryakov, hep-th/0003109.
63. I. Y. Aref'eva, arXiv:0912.5481 [hep-th].
64. S. S. Gubser, S. S. Pufu and A. Yarom, *Phys. Rev.* D **78**, 066014 (2008), arXiv:0805.1551.
65. S. S. Gubser, S. S. Pufu and A. Yarom, arXiv:0902.4062 [hep-th].
66. L. Alvarez-Gaume *et al.*, arXiv:0811.3969.
67. S. Lin and E. Shuryak, arXiv:0902.1508 [hep-th].
68. I. Y. Aref'eva, A. A. Bagrov and E. A. Guseva, *JEPH* 1130709 (2009) arXiv:0905.1087 [hep-th].
69. I. Y. Aref'eva, A. A. Bagrov and L. V. Joukovskaya, arXiv:0909.1294 [hep-th].
70. I. Y. Aref'eva, A. A. Bagrov and E. O. Pozdeeva, arXiv:1201.6542 [hep-th].
71. E. Kiritsis and A. Taliotis, arXiv:1111.1931 [hep-ph].

Chapter 11

SINE–GORDON THEORY IN THE REPULSIVE REGIME, THERMODYNAMIC BETHE ANSATZ AND MINIMAL MODELS

H. ITOYAMA

Department of Mathematics and Physics, Graduate School of Science,
Osaka City University,
Osaka City University Advanced Mathematical Institute (OCAMI)

3-3-138, Sugimoto, Sumiyoshi-ku, Osaka 558-8585, Japan

Neutral excitations present in the repulsive regime ($1/2 < \beta^2/8\pi < 1$) of the sine–Gordon/massive–Thirring model and its study of the massless limit by the thermodynamic Bethe ansatz is revisited. At $\beta^2/8\pi = 1 - 1/(p+1)$ the solitons become infinitely heavy, forcing truncation to the neutral excitations alone. The central charge in this limit is calculated to be $c = 1 - 6/p(p+1)$; the mass and S-matrices of the truncated theories are identified as those of the minimal conformal theory M_p perturbed by the $\phi_{(1,3)}$ operator.[a]

Keywords: Sine–Gordon model; massive–Thirring model; Bethe ansatz; Yang–Yang functional; minimal conformal series; Yang–Baxter relation.

1. Introduction

Professor Vladimir Korepin's series of work on the treatment of the sine–Gordon/massive–Thirring theory by the Bethe–ansatz/quantum inverse scattering in late seventies through early eighties has been a fascinating place for relativistic field theorists to learn the exactly diagonalizable quantum field theories and still serve as valuable references today. In this festshrift, I would like to recall my thought on the repulsive regime of the sine–Gordon theory, in particular, the issue of the universality classes,[1,2] which occasionally influenced my thought[3–5] in later years.

The mass spectrum and S-matrix of the sine–Gordon or equivalently the massive–Thirring model are well-known in the attractive regime.[6–11] These can be understood both from the Bethe ansatz diagonalization or quantum inverse scattering[12–14] and from the bootstrap program.[8] The equivalence of the sine–Gordon and massive–Thirring theories can be understood from the point of view of perturbation

---

[a]Contribution to the Vladimir Korepin festshrift.

theory.[15,16] Less well-known are the structure of the vacuum, the excitations, and the S-matrix in the repulsive regime.[17] The vacuum constructed just by filling in the negative-energy pseudoparticle excitations over the reference state is unstable for $\beta^2/8\pi \geq 2/3$. This is reminiscent of the level-crossing transitions that have received attention recently.[18–20] One must also fill the vacuum with the bound states known as string solutions in order to construct the true ground state. In other words, the vacuum consists of a multi-component condensate. (This fact has been confirmed using several different regularizations.[21])

Some time ago, it was found by Korepin[17] that the excitations built over this true vacuum include not only the solitons and antisolitons but also neutral excitations which appear when holes are made in the condensate. The masses and S-matrix have been computed. We will just quote the results due to Korepin.

The mass spectrum on the segment

$$1 - \frac{1}{\ell+1} < \frac{\beta^2}{8\pi} < 1 - \frac{1}{\ell+2}, \quad \ell = 1,2,\ldots \tag{1}$$

consists of $(\ell-1)$ neutral particles with masses

$$m_a = 2m \sin a\pi(1 - \beta^2/8\pi)/\tan\pi(1 - \beta^2/8\pi), \quad a = 1,\ldots,\ell-1 \tag{2}$$

and soliton and antisoliton of mass

$$m_s = m \left\{ \frac{\sin(\ell-1)\pi(1 - \beta^2/8\pi)}{\sin\pi(1 - \beta^2/8\pi)} \right.$$
$$\left. + \frac{\sin \ell\pi(1 - \beta^2/8\pi)}{\sin\pi(1 - \beta^2/8\pi)} \tan\frac{\pi}{2}\left(\frac{1}{1 - \beta^2/8\pi} - \ell - 1\right) \right\}. \tag{3}$$

Here m is related to the bare mass m_0 by a renormalization which we discuss below.

Here we shall discuss the critical properties of this system at the first-order phase transition point $\beta^2/8\pi = 1 - 1/(\ell+2)$,[1,2] where $\ell = p-1$, using the thermodynamic Bethe ansatz.[22] We first briefly review the formalism of the thermodynamic Bethe ansatz.[23,24] Consider N interacting particles that are built over the filled vacuum, so that their energy is positive definite. Of these particles, let N_a belong to species $a = 1,\ldots,n$. As there is no exchange of quantum number in the case of diagonal S-matrices, the periodic boundary conditions of a box with size L for a generic N-body state imply

$$e^{iLm_i \sinh v_i} \prod_{j\neq i} S_{ij}(v_i - v_j) = 1, \quad i,j = 1,\ldots,N. \tag{4}$$

Here S_{ij} is the two-body S-matrix, and m_i and v_i are respectively the physical (renormalized) masses and rapidities of the ith particle. In what follows we shall need several pieces of information arising from a field-theoretic context. In particular, we will introduce a rapidity cutoff $v_{\max}$ whenever required.

The goal is to determine the thermodynamic quantities from the distributions of particles $\rho_r^{(a)}$ and holes $\rho_h^{(a)}$ (the solutions of (4) not occupied by the particles), obeying the equation

$$\rho^{(a)}(v) = \frac{m_a}{2\pi}\cosh v + \sum_{b=1}^{n}\int_{-\infty}^{\infty}\frac{dv'}{2\pi}\varphi_{ab}(v-v')\rho_r^{(b)}(v') \tag{5}$$

derived from (4) in the limit $N \to \infty$, $L \to \infty$, $N/L =$ finite. We will assume that the states obey an exclusion principle. Here $\rho^{(a)} \equiv \rho_r^{(a)} + \rho_h^{(a)}$, and φ_{ab} is given by:

$$\varphi_{ab}(v) = -i\frac{d}{dv}\ln S_{ab}(v). \tag{6}$$

(S_{ab} is the S-matrix for scattering of a particle of type a and a particle of type b.)

We want to determine the most probable distributions $\rho_r^{(a)}$, $\rho_h^{(a)}$ as functions of the temperature T, which will dominate the expressions for the thermodynamic quantities in the infinite-volume limit. To do so we need to minimize the free energy per unit length $F = E - TS$ using the variational principle. Introducing $\varepsilon_a(v)$ by:

$$\frac{\rho_r^{(a)}(v)}{\rho^{(a)}(v)} \equiv \frac{1}{1+e^{\varepsilon_a(v)}} \equiv f(\varepsilon_a(v)), \tag{7}$$

we obtain

$$\frac{m_a}{T}\cosh v = \varepsilon_a(v) + \sum_{b=1}^{n}\int_{-\infty}^{\infty}\frac{dv'}{2\pi}\varphi_{ab}(v-v')\ln(1+e^{-\varepsilon_b(v')}). \tag{8}$$

The free energy (the leading correction to the vacuum energy) and the entropy at the minimum are:

$$F = -T^2\sum_{a=1}^{n}\int_{-\infty}^{\infty}\frac{dv}{2\pi}\ln\left(1+e^{-\varepsilon_a(v)}\right)\frac{m_a}{T}\cosh v, \tag{9}$$

$$S = \sum_{a=1}^{n}\int_{-\infty}^{\infty}dv\,\rho_r^{(a)}(v)\left[\varepsilon_a(v) + \left(1+e^{\varepsilon_a(v)}\right)\ln\left(1+e^{-\varepsilon_a(v)}\right)\right]. \tag{10}$$

Let us take the massless limit $r \equiv m/T \to 0$, keeping T fixed. Since the expression for the entropy contains a scale m, it vanishes as long as the rapidity is finite. The only nonvanishing contribution, therefore, comes from the region of integration $|v| \to \infty$. Replacing $\rho_r^{(a)}(v)$ in the integrand of the entropy by its asymptotic expression [obtained from (5) and (7)], and

$$S = \frac{\pi\tilde{c}(r)}{3}T, \tag{11}$$

one can derive[b] the following formula for $\tilde{c} \equiv \lim_{r\to 0}\tilde{c}(r)$:

$$\tilde{c} = \frac{6}{\pi^2}\sum_{a=1}^{n}\left(L[f(\varepsilon_a(0))] - L[f(\varepsilon_a(\infty))]\right), \tag{12}$$

---

[b]There is the contribution from $\varepsilon_a(\infty)$ which was originally overlooked.

where $\tilde{c} \equiv c - 12 d_{\min}$ with c the central charge and $d_{\min}$ the lowest scaling dimension, and the function $L(x)$ is Rogers' dilogarithm,

$$L(x) = -\frac{1}{2} \int_0^x dy \left[\frac{\ln y}{1-y} + \frac{\ln(1-y)}{y} \right]. \tag{13}$$

The set of numbers $\varepsilon_a(0)$ and $\varepsilon_a(\infty)$ are determined from (8) in the limits $r \to 0$, $v \to 0$ and $r \to 0$, $v \to \infty$, respectively. Precisely how these double limits must be taken is discussed below.

In the context of the sine–Gordon theory, (12) obviously applies to the class of reflectionless points $\beta^2/8\pi = 1/n$, $n = 2, 3, \ldots$. In fact, the calculation of $\tilde{c}$ is identical to the calculation for D_n factorized S-matrices. Under the assumption that $\varepsilon_a(\infty) = +\infty$ for $a = 1, \ldots, n-2$, soliton, antisoliton, we confirm that $\tilde{c} = 1$.

What is perhaps not so obvious is that the above formalism extends almost trivially to the class of nonreflectionless points $\beta^2/8\pi = 2/(2n+3)$, $n = 1, 2, \ldots$. Let us see how this is possible. In general, in the presence of exchange of quantum number, the set of scalar Eq. (4) is generalized to a set of coupled matrix equations,

$$e^{iLm_i \sinh v_i} \prod_{j \neq i} \mathbf{S}_{ij}(v_i - v_j) = 1. \tag{14}$$

Here, the matrix $\mathbf{S}_{ij}$ now carries two sets of indices, and the matrix multiplication is taken with respect to one set of indices. The soliton–antisoliton S-matrix can be identified mathematically with the Boltzmann weights of the six-vertex model having quantum-group parameter $q = -\exp(\pi i/\gamma')$, where $\gamma' = \beta^2/8\pi(1 - \beta^2/8\pi)$. At a generic value of the coupling constant, the diagonalization of the above equation is a formidable problem. In the case where q is a root of unity, however, the situation can be simplified by changing from the vertex basis to the path basis. This is done by using a q-Clebsch–Gordon coefficient as intertwiner.[25] When $q = \exp(\pi i/k)$, the "height" indices in the pat basis can take integer values $\ell, 0 \leq \ell \leq k - 1$. When $q = i$, there is no admissible path.

Thus, the soliton–antisoliton degree of freedom does not develop into a contribution to (14), and disappears when the infinite-volume limit is taken. This is a simple way to see the conclusion reached in Ref. 26. In fact, we find from (12) that $\tilde{c} = 2n/(2n+3)$. (We assume here that $\varepsilon_a(\infty) = +\infty$.)

Armed with these insights, we turn to the first-order phase transition points in the repulsive regime. Let us first list the S-matrices among solitons, anti-solitons and neutral particles.[17] In the fixed-parity basis,

$$S_{ss}\left(v \left| \frac{\beta^2}{8\pi} \right. \right) = S_Z^{\pm}\left(v \left| \left(\frac{\beta^2}{8\pi} \right)_{\ell} \right. \right),$$

$$S_Z^{\pm}\left(v \left| \frac{\beta^2}{8\pi} \right. \right) = U_{\pm}\left(v \left| \frac{\beta^2}{8\pi} \right. \right) S\left(v \left| \frac{\beta^2}{8\pi} \right. \right),$$

$$U_+\left(v \left| \frac{\beta^2}{8\pi} \right. \right) = \frac{\sinh(i\pi + v)/2\gamma'}{\sinh(i\pi - v)/2\gamma'}, \tag{15}$$

$$U_-\left(v\left|\frac{\beta^2}{8\pi}\right.\right) = -\frac{\cosh(v+i\pi)/2\gamma'}{\cosh(v-i\pi)/2\gamma'},$$

$$S\left(v\left|\frac{\beta^2}{8\pi}\right.\right) = \exp\left\{-\int_0^\infty \frac{dx}{x} \frac{\sinh((x/2\gamma')-x/2)\sinh(ivx/\pi\gamma')}{\sinh(x/2)\cosh(x/2\gamma')}\right\}$$

$$S_{aa'}(v) = \left(\frac{i\exp(v)+1}{\exp(v)+i}\right), \quad a'=a\pm 1,$$

$$S_{aa'}(v) = 1, \quad a'\neq a\pm 1,$$

$$(16)$$

$$S_{as}(v) = \left(\frac{i\exp(v)+1}{\exp(v)+i}\right), \quad a=\ell-1,$$

$$S_{as}(v) = 1, \quad a\neq\ell-1,$$

$$(17)$$

in the segment (1). Here,

$$\left(\frac{\beta^2}{8\pi}\right)_\ell \equiv \frac{1-\ell(1-\beta^2/8\pi)}{1-(\ell-1)(1-\beta^2/8\pi)}; \tag{18}$$

i.e., due to the nontrivial vacuum structure the argument of the Zamolodchikov S-matrix in this regime is modified so that $1/2 \le (\beta^2/8\pi)_\ell \le 2/3$.

We shall later make use of the following expressions for the multiplicative mass renormalization:

$$m_0 = Z_m m,$$

$$Z_m = \exp(A(\beta^2/8\pi)v_{\max}),$$

$$A(\beta^2/8\pi) = -1+\frac{\beta^2}{4\pi},$$

$$(19)$$

which follow from the consistency of the Bethe ansatz diagonalization.[9–11,17] Here $v_{\max} \approx \ln(\Lambda/\mu)$, where Λ is the ordinary momentum cutoff in relativistic field theory. This is a universal formula valid in both the attractive and repulsive regimes. The quantity A is just the anomalous dimension

$$\gamma_m = -\mu\frac{d}{d\mu}\ln Z_m = A(\beta^2/8\pi). \tag{20}$$

The crucial observation is that, in each segment, the points of interest give $(\beta^2/8\pi)_\ell = 2/3$. But this is simply the $n=0$ case of the cases considered above. By the same reasoning, the solitons and antisolitons fail to give a contribution to (14) and so disappear from the theory in the infinite-volume limit.

This fact can be seen more easily from the spectrum (2), (3). At the points under discussion, the soliton mass becomes infinite, decoupling from the theory. We are left with just the neutral excitations which do not have poles in the physical sheet. Let us now analyze the contribution of the neutral particles to the central charge in the massless limit, using (12).

We first discuss the limit $r \to 0$, $v \to 0$ in (8). A general feature of the solution $\varepsilon_a(v)$ of (8) is that it has a plateau of length $\sim \ln(1/r)$ around the origin. In this case, we can determine $\varepsilon_a(0)$ irrespective of how the limit is taken. The left-hand side of (8) can be neglected if we take $r \to 0$ followed by $v \to 0$. Defining

$$N_{ab} = -\frac{1}{2\pi} \int_{-\infty}^{\infty} dv \varphi_{ab}(v), \tag{21}$$

we find that $\varepsilon_a(0)$ satisfies the algebraic equations

$$e^{\varepsilon_a(0)} = \prod_{b=1}^{\ell-1} \left(1 + e^{-\varepsilon_a(0)}\right)^{N_{ab}}, \quad a = 1, \ldots, \ell-1, \tag{22}$$

where in our case N_{ab} is a tridiagonal $(\ell - 1) \times (\ell - 1)$ matrix with $-(1/2)$ in all nonvanishing entries:

$$N_{ab} = \begin{pmatrix} 0 & -\dfrac{1}{2} & 0 & \cdots & 0 & 0 & 0 \\[2mm] -\dfrac{1}{2} & 0 & -\dfrac{1}{2} & \cdots & 0 & 0 & 0 \\[2mm] 0 & -\dfrac{1}{2} & 0 & \cdots & 0 & 0 & 0 \\[2mm] \vdots & \vdots & \vdots & \ddots & \vdots & \vdots & \vdots \\[2mm] 0 & 0 & 0 & \cdots & 0 & -\dfrac{1}{2} & 0 \\[2mm] 0 & 0 & 0 & \cdots & -\dfrac{1}{2} & 0 & -\dfrac{1}{2} \\[2mm] 0 & 0 & 0 & \cdots & 0 & -\dfrac{1}{2} & 0 \end{pmatrix} \tag{23}$$

The solution to (22) is given by:

$$f(\varepsilon_a(0)) = 1 - \left(\frac{\sin \dfrac{\pi}{\ell + 2}}{\sin \dfrac{(a+1)\pi}{\ell + 2}}\right)^2, \quad a = 1, \ldots, \ell - 1. \tag{24}$$

The other limit $r \to 0$, $v \to \infty$ is much more delicate since the left-hand side of (8) is ambiguous unless a definite limiting procedure is specified. It is legitimate to send the renormalized mass to zero in accordance with (19) for fixed m_0, namely, $m(v_{\max}) = m_0 \exp[-A(\beta^2/8\pi)v_{\max}]$ with $v_{\max} \to \infty$; then we take $v \to \infty$ in (8). For the first $\ell-2$ neutral excitations, this discussion is sufficient. For the last neutral excitation, we have to take into account the fact that it still couples [cf. (17)] to the solitons whose masses have become infinite. From formula (30) of Ref. 17, which is used to deduce the cutoff dependence of the bare masses, it can be seen that only this last neutral excitation couples to the (infinite) vacuum density of the solitons. The Bethe ansatz equation then requires an additional subtraction of infinity, and the mass parameter for the last neutral excitation fails to be multiplicatively renormalizable:

$$m_{\ell-1} = Z_m^{-1} m_0 + \delta m \, ; \qquad (25)$$

therefore, when we take $v_{\max} \to \infty$ the renormalized mass stays finite and the left-hand side of (8) goes to infinity as $v \to \infty$. Thus $\varepsilon_a(v)$ for the last neutral particle goes to infinity at a distance $\sim \ln(1/r)$, while $\varepsilon_a(v)$ for the others remains finite.

Putting together all these results, we find that $\varepsilon_a(\infty)$ obeys:

$$e^{\varepsilon_a(\infty)} = \prod_{b=1}^{\ell-2} \left(1 + e^{-\varepsilon_a(\infty)}\right)^{N_{ab}} , \qquad a = 1, \ldots, \ell-2 , \qquad (26)$$

$$e^{\varepsilon_{\ell-1}(\infty)} = +\infty \, .$$

The solution to (26) is given by:

$$f(\varepsilon_a(\infty)) = 1 - \left(\frac{\sin \dfrac{\pi}{\ell+1}}{\sin \dfrac{(a+1)\pi}{\ell+1}} \right)^2 , \qquad a = 1, \ldots, \ell-2 . \qquad (27)$$

From (12), (24) and (27) we obtain the central charge for the minimal conformal series

$$\tilde{c} = 1 - \frac{6}{(\ell+1)(\ell+2)} = 1 - \frac{6}{p(p+1)} \, . \qquad (28)$$

In obtaining this equation, we have used $L(x) = L(1) - L(1-x)$, $L(1) = \pi^2/6$, as well as the summation formula

$$\sum_{a=1}^{k} L \left(\frac{\sin^2 \left(\dfrac{\pi}{k+3} \right)}{\sin^2 \left(\dfrac{\pi(a+1)}{k+3} \right)} \right) = \frac{2k}{k+3} L(1) , \qquad (29)$$

for k an integer ≥ 1.

We also observe that the anomalous dimension γ_m computed from (19) is

$$\gamma_m = \frac{\ell}{\ell+2} , \qquad (30)$$

in agreement with the highest weight $h_{1,3}$ of the minimal conformal theory M_p. These results provide a complete demonstration that the points $\beta^2/8\pi = 1 - 1/(\ell+2)$, at which truncation to the neutral excitations automatically occurs, represent infrared stable (i.e., stable as the cutoff goes to infinity) massive theories obtained from the M_p theories by perturbing with the $\phi_{(1,3)}$ operator. We find that, in general,

$$\frac{\beta^2}{8\pi} = \frac{p}{p'} \Rightarrow \gamma_m = h_{1,3} \text{ of } M_{p/p'} , \qquad (31)$$

where p and p' are coprime integers.

Let us come back to the attractive regime. One must push the bare mass to zero in order to tale the limit. If we take m to be constant when sending $v_{\max} \to \infty$, followed by $v \to \infty$ in (8), then we find $\varepsilon_a(+\infty) = +\infty$, confirming the assumption which led to $\tilde{c} = 1$ in this regime.[24]

Finally, let us discuss the consistency of our results with the existing literature. First, we find a striking resemblance between our discussion of the massless limit of the sine–Gordon theory based on the thermodynamic Bethe ansatz and the calculation by Bazhanov and Reshetikhin[27] in the critical restricted solid-on-solid models. In particular, the same tridiagonal matrix (23) appears in their work. This evidence has convinced us that the properties (common to several regularization schemes) of the theory treated here are in fact the same as those of the massive scaling limit of the model first considered by Andrews, Baxter and Forrester[28] in regime III — it is an explicit integrable realization of the minimal models away from criticality.[29] Our result that the anomalous dimension is not modified away from criticality points towards the more general phenomenon of a noncritical Virasoro structure revealed in Ref. 30 and 31.

References

1. H. Itoyama and P. Moxhay, *Phys. Rev. Lett.* **65**, 2102 (1990).
2. COLO-HEP, ITP-SB-90-64, June 1990, in Argonne 1990, Proceedings, Quantum groups 306-320.
3. H. Itoyama and T. Oota, *Nucl. Phys. B* **419**, 632 (1994).
4. H. Itoyama and T. Oota, *Prog. Theor. Phys. Suppl.* **114**, 41 (1993).
5. A. Fujii and H. Itoyama, *Phys. Rev. Lett.* **86**, 5235 (2001).
6. J. Johnson, S. Krinsky and B. McCoy, *Phys. Rev. A* **8**, 526 (1973).
7. A. Luther, *Phys. Rev. B* **14**, 2153 (1975).
8. A. B. Zamolodchikov and Al. B. Zamolodchikov, *Ann. Phys.* **120**, 253 (1979).
9. H. Bergknoff and H. B. Thacker, *Phys. Rev. Lett.* **42**, 135 (1979).
10. H. Bergknoff and H. B. Thacker, *Phys. Rev. D* **19**, 3666 (1979).
11. V. E. Korepin, *Teor. Mat. Fiz.* **41**, 169 (1979) [*Theor. Math. Phys.* **41**, 953 (1979)].
12. L. D. Faddeev, in *Soviet Scientific Reviews*, Vol. 1, ed. S. P. Novikov (Harwood Academic, New York, 1980), Sect. C, p. 107.
13. H. B. Thacker, *Rev. Mod. Phys.* **53**, 253 (1981).
14. P. P. Kulish and E. K. Sklyanin, in *Integrable Quantum Field Theories*, eds. J. Hietarinta and C. Montonen (Springer-Verlag, New York, 1982).
15. S. Coleman, *Phys. Rev. D* **11**, 2088 (1975).
16. S. Mandelstam, *Phys. Rev. D* **11**, 3026 (1975).
17. V. E. Korepin, *Commun. Math. Phys.* **76**, 165 (1980).
18. G. Albertini, B. M. McCoy and J. H. H. Perk, *Phys. Lett. A* **135**, 159 (1989).
19. G. Albertini, B. M. McCoy and J. H. H. Perk, *Phys. Lett. A* **139**, 204 (1989).
20. H. Itoyama, B. M. McCoy and J. H. H. Perk, *Int. J. Mod. Phys. B* **4**, 995 (1990).
21. N. M. Bogoliubov, V. E. Korepin and A. G. Izergin, *Phys. Lett. B* **159**, 345 (1985).
22. C. N. Yang and C. P. Yang, *J. Math. Phys.* **10**, 1115 (1969).
23. Al. B. Zamolodchikov, *Nucl. Phys. B* **342**, 695 (1990).
24. T. R. Klassen and E. Melzer, *Nucl. Phys. B* **338**, 485 (1990).

25. V. Pasquier, *Commun. Math. Phys.* **118**, 355 (1988).
26. F. A. Smirnov *Int J. Mod. Phys. A* **4**, 4213 (1989).
27. V. V. Bazhanov and N. Yu. Reshetikhin, *Int. J. Mod. Phys. A* **4**, 115 (1989).
28. G. E. Andrews, R. J. Baxter and P. J. Forrester, *J. Stat. Phys.* **35**, 193 (1984).
29. D. A. Huse, *Phys. Rev. B* **30**, 3908 (1984).
30. H. Itoyama and H. B. Thacker, *Phys. Rev. Lett.* **58**, 1395 (1987).
31. H. Itoyama and H. B. Thacker, *Nucl. Phys. B* **320**, 541 (1989).
32. H. Itoyama and H. B. Thacker, *Nucl. Phys. B (Proc. Suppl.)* **5A**, 9 (1988).

Chapter 12

ON SOME ALGEBRAIC AND COMBINATORIAL
PROPERTIES OF DUNKL ELEMENTS

ANATOL N. KIRILLOV

Research Institute of Mathematical Sciences (RIMS),
Kyoto 606-8502, Japan
kirillov@kurims.kyoto-u.ac.jp

To Volodya Korepin, my friend and co-author,

on the occasion of his sixtieth birthday, with grateful and admiration

We introduce and study a certain class of nonhomogeneous quadratic algebras together with the special set of mutually commuting elements inside of each, the so-called *Dunkl elements*. We describe relations among the Dunkl elements. This result is a further generalization of similar results obtained in [S. Fomin and A. N. Kirillov, Quadratic algebras, Dunkl elements and Schubert calculus, in *Advances in Geometry* (eds. J.-S. Brylinski, V. Nistor, B. Tsygan and P. Xu), *Progress in Math.* Vol. 172 (Birkhäuser Boston, Boston, 1995), pp. 147–182, A. Postnikov, On a quantum version of Pieri's formula, in *Advances in Geometry* (eds. J.-S. Brylinski, R. Brylinski, V. Nistor, B. Tsygan and P. Xu), *Progress in Math.* Vol. 172 (Birkhäuser Boston, 1995), pp. 371–383 and A. N. Kirillov and T. Maenor, *A Note on Quantum K-Theory of Flag Varieties*, preprint]. As an application we describe explicitly the set of relations among the Gaudin elements in the group ring of the symmetric group, cf. [E. Mukhin, V. Tarasov and A. Varchenko, Bethe Subalgebras of the Group Algebra of the Symmetric Group, preprint arXiv:1004.4248].

Also we describe a few combinatorial properties of some special elements in the associative quasi-classical Yang–Baxter algebra in a connection with the values of the β-Grothendieck polynomials for some special permutations, and on the other hand, with the Ehrhart polynomial of the Chan–Robbins polytope.

Keywords: Dunkl and Gaudin operators at critical level; Catalan and Schroder numbers; Schubert and Grothendieck polynomials.

1. Introduction

The Dunkl operators have been introduced in the later part of 80's of the last century by Charles Dunkl[4,5] as a powerful mean to study of harmonic and orthogonal polynomials related with finite Coxeter groups. In the present paper we do not need the definition of Dunkl operators for arbitrary (finite) Coxeter groups, see e.g., Ref. 4, but only for the special case of the symmetric group $\mathbb{S}_n$.

Definition 1.1. Let $P_n = \mathbb{C}[x_1, \ldots, x_n]$ be the ring of polynomials in variables

$x_1, \ldots, x_n$. The type A_{n-1} (additive) rational Dunkl operators $D_1, \ldots, D_n$ are the differential-difference operators of the following form

$$D_i = \lambda \frac{\partial}{\partial x_i} + \sum_{j \neq i} \frac{1 - s_{ij}}{x_i - x_j}. \tag{1}$$

Here s_{ij}, $1 \leq i < j \leq n$, denotes the exchange (or permutation) operator, namely,

$$s_{ij}(f)(x_1, \ldots, x_i, \ldots, x_j, \ldots, x_n) = f(x_1, \ldots, x_j, \ldots, x_i, \ldots, x_n);$$

$\partial / \partial x_i$ stands for the derivative w.r.t. the variable x_i; $\lambda \in \mathbb{C}$ is a parameter.

The key property of the Dunkl operators is the following result.

Theorem 1.1. (Dunkl,[4]) *For any finite Coxeter group* (W, S)*, where* $S = \{s_1, \ldots, s_l\}$ *denotes the set of simple reflections, the Dunkl operators* $D_i := D_{s_i}$ *and* $D_j := D_{s_j}$ *commute:*

$$D_i D_j = D_j D_i, \quad 1 \leq i, \quad j \leq l.$$

Another fundamental property of the Dunkl operators which finds wide variety of applications in the theory of integrable systems, see e.g., Ref. 12, is the following statement: operator

$$\sum_{i=1}^{l} (D_i)^2$$

"essentially" coincides with the Hamiltonian of the rational Calogero–Moser model related to the finite Coxeter group (W, S).

Definition 1.2. Truncated (additive) Dunkl operator (or the Dunkl operator at critical level), denoted by $\mathcal{D}_i$, $i = 1, \ldots, l$, is an operator of the form (1) with parameter $\lambda = 0$.

For example, the type A_{n-1} rational truncated Dunkl operator has the following form

$$\mathcal{D}_i = \sum_{j \neq i} \frac{1 - s_{ij}}{x_i - x_j}.$$

Clearly the truncated Dunkl operators generate a commutative algebra.

The important property of the truncated Dunkl operators is the following result discovered and proved by Dunkl[5]; see also Ref. 1 for a more recent proof.

Theorem 1.2. (Dunkl,[5] Bazlov[1]) *For any finite Coxeter group* (W, S) *the algebra over* $\mathbb{Q}$ *generated by the truncated Dunkl operators* $\mathcal{D}_1, \ldots, \mathcal{D}_l$ *is canonically isomorphic to the coinvariant algebra of the Coxeter group* (W, S)*.*

Example 1.1. In the case when $W = \mathbb{S}_n$ is the symmetric group, Theorem 1.2 states that the algebra over $\mathbb{Q}$ generated by the truncated Dunkl operators $\mathcal{D}_i = \sum_{j \neq i}(1 - s_{ij})/(x_i - x_j)$, $i = 1, \ldots, n$, is canonically isomorphic to the cohomology ring of the full flag variety $\mathcal{F}l_n$ of type A_{n-1}

$$\mathbb{Q}[\mathcal{D}_1, \ldots, \mathcal{D}_n] \cong \mathbb{Q}[x_1, \ldots, x_n]/J_n \,, \tag{2}$$

where J_n denotes the ideal generated by elementary symmetric polynomials $\{e_k(X_n), 1 \leq k \leq n\}$.

Recall that the symmetric polynomials $e_i(X_n)$, $i = 1, \ldots, n$, are defined through the generating function

$$1 + \sum_{i=1}^{n} e_i(X_n \, t^i = \prod_{i=1}^{n}(1 + t \, x_i) \,,$$

where we set $X_n := (x_1, \ldots, x_n)$. It is well-known that in the case $W = \mathbb{S}_n$, the isomorphism (2) can be defined over the ring of integers $\mathbb{Z}$.

Theorem 1.2 by Dunkl has raised a number of questions:

(**A**) What is the algebra generated by the truncated

- trigonometric,
- elliptic,
- super, matrix, $\ldots$,

    (a) additive Dunkl operators ?
    (b) Ruijsenaars–Schneider–Macdonald operators ?
    (c) Gaudin operators ?

(**B**) Describe commutative subalgebra generated by the Jucys–Murphy elements in

- the group ring of the symmetric group;
- the Hecke algebra;
- the Brauer algebra, BMV algebra, $\ldots$.

(**C**) Does there exist an analogue of Theorem 1.2 for

- Classical and quantum equivariant cohomology and equivariant K-theory rings of the flag varieties ?
- Cohomology and K-theory rings of affine flag varieties ?
- Diagonal coinvariant algebras of finite Coxeter groups ?
- Complex reflection groups ?

The present paper is a short Introduction to a few items from Sec. 5 of Ref. 14.

The main purpose of my paper "On some quadratic algebras, II" is to give some partial answers on the above questions in the case of the symmetric group $\mathbb{S}_n$.

The purpose of the <u>present paper</u> is to draw attention to an interesting class of nonhomogeneous quadratic algebras closely connected (still mysteriously!) with different branches of Mathematics such as:

Classical and Quantum Schubert and Grothendieck Calculi,
Low dimensional Topology,
Classical, Basic and Elliptic Hypergeometric functions,
Algebraic Combinatorics and Graph Theory,
Integrable Systems,
. . . .

What we try to explain in Ref. 14 is that after passing to *a suitable represen-tation* of the quadratic algebra in question, the subjects mentioned above, are a manifestation of certain general properties of that quadratic algebra.

From this point of view, we treat the commutative subalgebra generated by the additive (resp. multiplicative) truncated Dunkl elements in the algebra $3T_n(\beta)$, see Definition 3.1, as *universal cohomology* (resp. *universal K-theory*) ring of the flag variety $\mathcal{F}l_n$. The classical or quantum cohomology (resp. the classical or quantum K-theory) rings of $\mathcal{F}l_n$ are certain quotients of the *universal ring.*

For example, in Ref. 16 we have computed relations among the (truncated) Dunkl elements $\{\theta_i, i = 1, \ldots, n\}$ in the *elliptic representation* of the algebra $3T_n$ ($\beta = 0$). We expect that the commutative algebra obtained is isomorphic to (yet not defined, but see Ref. 11) the *elliptic cohomology* of flag variety $\mathcal{F}l_n$.

Another example from Ref. 14. Consider the algebra $3T_n$ ($\beta = 0$).

One can prove the following *identities* in the algebra $3T_n$ ($\beta = 0$).

(A) Summation formula

$$\sum_{j=1}^{n-1} \left(\prod_{b=j+1}^{n-1} u_{b,b+1} \right) u_{1,n} \left(\prod_{b=1}^{j-1} u_{b,b+1} \right) = \prod_{a=1}^{n-1} u_{a,a+1}.$$

(B) Duality transformation formula Let $m \leq n$, then

$$\sum_{j=m}^{n-1} \left(\prod_{b=j+1}^{n-1} u_{b,b+1} \right) \left[\prod_{a=1}^{m-1} u_{a,a+n-1} u_{a,a+n} \right] u_{m,m+n-1} \left(\prod_{b=m}^{j-1} u_{b,b+1} \right)$$

$$= \sum_{j=1}^{m} \left[\prod_{a=1}^{m-j} u_{a,a+n} u_{a+1,a+n} \right] \left(\prod_{b=m}^{n-1} u_{b,b+1} \right) \left[\prod_{a=1}^{j-1} u_{a,a+n-1} u_{a,a+n} \right]$$

$$- \sum_{j=2}^{m} \left[\prod_{a=j}^{m-1} u_{a,a+n-1} u_{a,a+n} \right] u_{m,n+m-1} \left(\prod_{b=m}^{n-1} u_{b,b+1} \right) u_{1,n}.$$

One can check that after passing to the *elliptic representation* of the algebra $3T_n$ ($\beta = 0$), see Ref. 14, Sec. 5.1.7, or Ref. 16 for the definition of *elliptic representation*, the above identities (**A**) and (**B**) finally end up correspondingly, as the Summation formula and Duality transformation formula for multiple elliptic hypergeometric series (of type A_{n-1}), see e.g., Ref. 13 for definition of the latter.

After passing to the so-called *Fay representation*,[14] the identities (**A**) and (**B**) become correspondingly the Summation formula and Duality transformation formula for the Riemann theta functions of genus $g > 0$.[14] These formulas in the case $g \geq 2$ seems to be new.

A few words about the content of the present paper.

In Sec. 1, I introduce the so-called *dynamical classical Yang–Baxter algebra* as "a natural quadratic algebra" in which the Dunkl elements form a pair-wise commuting family. It is the study of the algebra generated by the (truncated) Dunkl elements that is the main objective of our investigation in Ref. 14 and the present paper.

In Sec. 2 we introduce the algebra $3HT_n$ which seems to be the most general (noncommutative) deformation of the (even) Orlik–Solomon algebra, such that it's still possible to describe relations among the Dunkl elements, see Theorem 3.1. As an application we describe explicitly a set of relations among the (additive) Gaudin/Dunkl elements, cf. Ref. 24.

In Sec. 3 we describe some combinatorial properties of special elements in the associative quasi-classical Yang–Baxter algebra. The results of Sec. 4.1, see Theorem 4.1, items (**1**)–(**5**), are more or less known among the specialists in the subject, while those of the item (**6**) seems to be new. In Sec. 4.2 we give a partial answer on the question **6.C8**(d) by Stanley.[28]

At the end of Introduction I want to add two remarks.

(a) After a suitable modification of the algebra $3HT_n$, see Ref. 18, and the case $\beta \neq 0$ in Ref. 14, one can compute the set of relations among the (additive) Dunkl elements [defined in Sec. 1, (1.3)]. In the case $\beta = 0$ and $q_{ij} = q_i \delta_{j-i,1}$, $1 \leq i < j \leq n$, where $\delta_{a,b}$ is the Kronecker delta, the commutative algebra generated by additive Dunkl elements (1.3) appears to be "almost" isomorphic to the equivariant quantum cohomology ring of the flag variety $\mathcal{F}l_n$, see Ref. 18 for details. Using the multiplicative version of Dunkl elements (1.3), one can extend the results from Ref. 18 to the case of eqivariant quantum K-theory of the flag variety $\mathcal{F}l_n$, see Ref. 14.

(b) In fact one can construct an analogue of the algebra $3HT_n$ and a commutative subalgebra inside it, for any graph $\Gamma = (V, E)$ without loops and multiple edges.[14] We denote this algebra by $3T_n(\Gamma)$, and denote by $3T_n^{(0)}(\Gamma)$ its *nil-quotient*, which corresponds to the "classical limit of the algebra $3T_n(\Gamma)$".

The case of the complete graph $\Gamma = K_n$ reproduces the results of the present paper and that of Ref. 14, i.e., the case of the full flag variety $\mathcal{F}l_n$. The case of the *complete multipartite graph* $\Gamma = K_{n_1,\ldots,n_r}$ reproduces the analogues of results stated for the full flag variety $\mathcal{F}l_n$, in the case of the partial flag variety $\mathcal{F}_{n_1,\ldots,n_r}$, see Ref. 14 for details.

Example 1.2. Take $\Gamma = K_{2,2}$. The algebra $3T^{(0)}(\Gamma)$ is generated by four elements $\{a = u_{13}, b = u_{14}, c = u_{23}, d = u_{24}\}$ subject to the following set of (defining) relations

- $a^2 = b^2 = c^2 = d^2 = 0$, $cb = bc$, $ad = da$,
- $aba + bab = 0 = aca + cac$, $bdb + dbd = 0 = cdc + dcd$,
- $abd - bdc - cab + dca = 0 = acd - bac - cdb + dba$,
- $abca + adbc + badb + bcad + cadc + dbcd = 0$.

It is not difficult to see that:

$$\text{Hilb}(3T^{(0)}(K_{2,2}), t) = [3]_t^2 [4]_t^2, \quad \text{Hilb}(3T^{(0)}(K_{2,2})^{ab}, t) = (1, 4, 6, 3).$$

Here for any algebra A we denote by A^{ab} its <u>abelization</u>.

The commutative subalgebra in $3T^{(0)}(K_{2,2})$, which corresponds to the intersection $3T^{(0)}(K_{2,2}) \cap \mathbb{Z}[\theta_1, \theta_2, \theta_3, \theta_4]$, is generated by the elements $c_1 := \theta_1 + \theta_2 = (a + b + c + d)$ and $c_2 := \theta_1 \theta_2 = (ac + ca + bd + db + ad + bc)$. The elements c_1 and c_2 commute and satisfy the following relations

$$c_1^3 - 2c_1 c_2 = 0, \quad c_2^2 - c_1^2 c_2 = 0.$$

The ring of polynomials $\mathbb{Z}[c_1, c_2]$ is isomorphic to the cohomology ring $H^*(Gr(2,4), \mathbb{Z})$ of the Grassmannian variety $Gr(2,4)$.

This example is illustrative of the similar results valid for the graph $K_{n_1,\dots,n_r}$, i.e., for partial flag varieties. The meaning of the algebra $3T_n^{(0)}(\Gamma)$ and the corresponding commutative subalgebra inside it for a general graph Γ, is still unclear.

Conjecture 1. *Let* $\Gamma = (V, E)$ *be a connected subgraph of the complete graph* K_n *on* n *vertices. Then*

$$\text{Hilb}(3T_n^{(0)}(\Gamma)^{ab}, t) = t^{|V|-1} T(\Gamma; 1 + t^{-1}, 0),$$

where for any graph Γ *the symbol* $T(\Gamma; x, y)$ *denotes the* **Tutte polynomial** *corresponding to this graph.*

2. Dunkl Elements

Let $\mathcal{A}_n$ be the free associative algebra over $\mathbb{Z}$ with the set of generators $\{u_{ij}, 1 \leq i, j \leq n\}$. We set $x_i := u_{ii}$, $i = 1, \dots, n$.

Definition 2.1. Define (additive) Dunkl elements θ_i, $i = 1, \dots, n$, in the algebra $\mathcal{A}_n$ to be:

$$\theta_i = x_i + \sum_{\substack{j=1 \\ j \neq i}}^{n} u_{ij}. \tag{3}$$

We are interested in to find "natural relations" among the generators $\{u_{ij}\}$ such that the Dunkl elements (3) are pair-wise commute. One natural condition which is the commonly accepted in the theory of integrable systems, is:

- (Locality condition)

$$[x_i, x_j] = 0, \quad u_{ij} u_{kl} = u_{kl} u_{ij}, \quad \text{if} \quad \{i, j\} \cap \{k, l\} = \emptyset. \tag{4}$$

Lemma 2.1. *Assume that elements $\{u_{ij}\}$ satisfy the locality conditions* (4). *Then*

$$[\theta_i, \theta_j] = \left[x_i + \sum_{k \neq i,j} u_{ik}, u_{ij} + u_{ji}\right] + \left[u_{ij}, \sum_{k=1}^{n} x_k\right] + \sum_{k \neq i,j} w_{ijk},$$

where

$$w_{ijk} = [u_{ij}, u_{ik} + u_{jk}] + [u_{ik}, u_{jk}] + [x_i, u_{jk}] + [u_{ik}, x_j] + [x_k, u_{ij}]. \tag{5}$$

Therefore in order to ensure that the Dunkl elements form a pair-wise commuting family, it's natural to assume that the following conditions hold

- (Unitarity)

$$[u_{ij} + u_{ji}, u_{kl}] = 0 = [u_{ij} + u_{ji}, x_k] \quad \text{for all } i, j, k, l, \tag{6}$$

i.e., the elements $u_{ij} + u_{ji}$ are <u>central</u>.

- (Crossing relations)

$$\left[\sum_{k=1}^{n} x_k, u_{ij}\right] = 0 \quad \text{for all } i, j. \tag{7}$$

- (Dynamical classical Yang–Baxter relations)

$$[u_{ij}, u_{ik} + u_{jk}] + [u_{ik}, u_{jk}] + [x_i, u_{jk}] + [u_{ik}, x_j] + [x_k, u_{ij}] = 0, \tag{8}$$

if i, j, k are pair-wise distinct.

We denote by $DCYB_n$ the quotient of the algebra $\mathcal{A}_n$ by the two-sided ideal generated by relations (4), (6), (7) and (8). Clearly, the Dunkl elements (3) generate a commutative subalgebra inside the algebra $CDYB_n$, and the sum $\sum_{i=1}^{n} \theta_i = \sum_{i=1}^{n} x_i$ belongs to the center of the algebra $DCYB_n$.

Example 2.1. (A representation of the algebra $DCYB_n$, cf Ref. 6).

Given a set $q_1, \ldots, q_{n-1}$ of mutually commuting parameters, define $q_{ij} = \prod_{a=i}^{j-1} q_a$, if $i < j$ and set $q_{ij} = q_{ji}$ in the case $i > j$. Clearly, that if $i < j < k$, then $q_{ij} q_{jk} = q_{ik}$.

Let $z_1, \ldots, z_n$ be a set of variables. Denote by $P_n := \mathbb{Z}[z_1, \ldots, z_n]$ the corresponding ring of polynomials. We consider the variable z_i, $i = 1, \ldots, n$, also as the operator acting on the ring of polynomials P_n by multiplication on z_i..

Let $s_{ij} \in \mathbb{S}_n$ be a transposition. We consider the transposition s_{ij} also as the operator which acts on the ring P_n by interchanging z_i and z_j, and fixes all other variables. We denote by:

$$\partial_{ij} = \frac{1 - s_{ij}}{z_i - z_j}$$

the divided difference operator corresponding to the transposition s_{ij}. Finally we define operator (cf. Ref. 6)

$$\partial_{(ij)} := \partial_i \cdots \partial_{j-1} \partial_j \partial_{j-1} \cdots \partial_i, \quad \text{if } i < j.$$

The operators $\partial_{(ij)}$, $1 \leq i < j \leq n$ satisfy (among others) the following set of relations (cf. Ref. 6)

- $[z_j, \partial_{(ik)}] = 0$, if $j \notin [i, k]$, $[\partial_{(ij)}, \sum_{a=i}^{j} z_a] = 0$,
- $[\partial_{(ij)}, \partial_{(kl)}] = \delta_{jk} [z_j, \partial_{(il)}] + \delta_{il} [\partial_{(kj)}, z_i]$, if $i < j$, $k < l$.

Therefore, if we set $u_{ij} = q_{ij}\partial_{(ij)}$, if $i < j$ and $u_{(ij)} = -u_{(ji)}$, if $i > j$, then for a triple $i < j < k$ we will have:

$$[u_{ij}, u_{ik} + u_{jk}] + [u_{ik}, u_{jk}] + [z_i, u_{jk}] + [u_{ik}, z_j] + [z_k, u_{jk}]$$
$$= q_{ij}q_{jk}[\partial_{(ij)}, \partial_{(jk)}] + q_{ik}[\partial_{(ik)}, z_j] = 0.$$

Thus the elements z_i, $i = 1, \ldots, n$ and $\{u_{ij}, 1 \leq i < j \leq n\}$ define a representation of the algebra $DCYB_n$, and therefore the Dunkl elements

$$\theta_i := z_i + \sum_{j \neq i} u_{ij} = z_i - \sum_{j < i} q_{ji}\partial_{(ji)} + \sum_{j > i} q_{ij}\partial_{(ij)}$$

form a pairwise commuting family of operators acting on the ring of polynomials $\mathbb{Z}[q_1, \ldots, q_{n-1}][z_1, \ldots, z_n]$, cf. Ref. 6.

Comments 2.1. (Nonunitary dynamical classical Yang–Baxter algebra) Let $\tilde{\mathcal{A}}_n$ be the quotient of the algebra $\mathcal{A}_n$ by the two-sided ideal generated by the relations (4), (7) and (8). Consider elements

$$\theta_i = x_i + \sum_{a \neq i} u_{ia}, \quad \text{and} \quad \bar{\theta}_j = -x_j + \sum_{b \neq j} u_{bj}, \quad 1 \leq i < j \leq n.$$

Then

$$[\theta_i, \bar{\theta}_j] = \left[\sum_{k=1}^{n} x_k, u_{ij}\right] + \sum_{k \neq i,j} w_{ikj}.$$

Therefore the elements θ_i and $\bar{\theta}_j$ commute in the algebra $\tilde{\mathcal{A}}_n$.

In the case when $x_i = 0$ for all $i = 1, \ldots, n$, the relations $w_{ijk} = 0$, assuming that i, j, k are all distinct, are well-known as the (nonunitary) classical Yang–Baxter relations. Note that for a given triple (i, j, k) we have in fact six relations. These six relations imply that $[\theta_i, \bar{\theta}_j] = 0$. However,

$$[\theta_i, \theta_j] = \left[\sum_{k \neq i,j} u_{ik}, u_{ij} + u_{ji}\right] \neq 0.$$

In order to ensure the commutativity relations among the Dunkl elements, i.e., $[\theta_i, \theta_j] = 0$ for all i, j, one needs to impose on the elements $\{u_{ij}, 1 \leq i \neq j \leq n\}$ the "twisted" classical Yang–Baxter relations, namely

$$[u_{ij} + u_{ik}, u_{jk}] + [u_{ik}, \mathbf{u}_{ji}] = 0, \quad \text{if} \quad i, j, k \text{ are all distinct}. \tag{9}$$

Contrary to the case of nonunitary classical Yang–Baxter relations, it is easy to see that in the case of twisted classical Yang–Baxter relations, for a given triple (i, j, k) one has only three relations.

3. Algebra $3HT_n$

Consider the twisted classical Yang–Baxter relation

$$[u_{ij} + u_{ia}, u_{ja}] + [u_{ia}, u_{ji}] = 0, \quad \text{where } i, j, k \text{ are distinct}.$$

Having in mind applications of the Dunkl elements, we split the above relation on two relations

$$\underline{u_{ij} u_{jk} = u_{jk} u_{ik} - u_{ik} u_{ji}} \quad \text{and} \quad \underline{u_{jk} u_{ij} = u_{ik} u_{jk} - u_{ji} u_{ik}},$$

and impose the unitarity constraints

$$u_{ij} + u_{ji} = \beta,$$

where β is a central element. Summarizing, we come to the following definition.

Definition 3.1. (Cf. Ref. 9 and 15) Define algebra $3T_n(\beta)$ to be the quotient of the free associative algebra $\mathbb{Z}[\beta] \langle u_{ij}, 1 \leq i < j \leq n \rangle$ by the set of relations

- (Locality) $u_{ij} u_{kl} = u_{kl} u_{ij}$, if $\{i, j\} \cap \{k, l\} = \emptyset$,
- $u_{ij} u_{jk} = u_{ik} u_{ij} + u_{jk} u_{ik} \beta u_{ik}$, $u_{jk} u_{ij} = u_{ij} u_{ik} + u_{ik} u_{jk} - \beta u_{ik}$, if $1 \leq i < j < k \leq n$.

For each pair $i < j$, we define element $q_{ij} := u_{ij}^2 - \beta u_{ij} \in 3T_n(\beta)$.

Lemma 3.1.
(1) *The elements $\{q_{ij}, 1 \leq i < j \leq n\}$ satisfy the Kohno–Drinfeld relations (known also as the horizontal four term relations)*

$$q_{ij} q_{kl} = q_{kl} q_{ij}, \quad if \quad \{i, j\} \cap \{k, l\} = \emptyset,$$

$$[q_{ij}, q_{ik} + q_{jk}] = 0, \quad [q_{ij} + q_{ik}, q_{jk}] = 0, \quad if \quad i < j < k.$$

(2) *For a triple $(i < j < k)$ define $u_{ijk} := u_{ij} - u_{ik} + u_{jk}$. Then*

$$u_{ijk}^2 = \beta u_{ijk} + q_{ij} + q_{ik} + q_{jk}$$

(3) *(Deviation from the Yang–Baxter and Coxeter relations)*

$$u_{ij} u_{ik} u_{jk} - u_{jk} u_{ik} u_{ij} = [u_{ik}, q_{ij}] = [q_{jk}, u_{ik}],$$

$$u_{ij} u_{jk} u_{ij} - u_{jk} u_{ij} u_{jk} = q_{ij} u_{ik} - u_{ik} q_{jk}.$$

Comments 3.1. It is easy to see that the horizontal four-term relations listed in Lemma 3.1, (1), are equivalent to the locality condition among the generators $\{q_{ij}\}$, together with the commutativity conditions among the Jucys–Murphy elements

$$d_i := \sum_{j=i+1}^{n} q_{ij}, \quad i = 2, \ldots, n,$$

namely, $[d_i, d_j] = 0$. In Ref. 14 we describe some properties of a commutative subalgebra generated by the Jucys–Murphy elements in the Kohno–Drinfeld algebra.

It is well-known that the Jucys–Murphy elements generate a maximal commutative subalgebra in the group ring of the symmetric group $\mathbb{S}_n$. It is an open problem to describe defining relations among the Jucys–Murphy elements in the group ring $\mathbb{Z}[\mathbb{S}_n]$.

Finally we introduce the "Hecke quotient" of the algebra $3T_n(\beta)$, denoted by $3HT_n$.

Definition 3.2. Define algebra $3HT_n$ to be the quotient of the algebra $3T_n(\beta)$ by the set of relations

$$q_{ij}q_{kl} = q_{kl}q_{ij}, \quad \text{for all } i,j,k,l.$$

In other words we assume that the all elements $\{q_{ij}, 1 \leq i < j \leq n\}$ are central in the algebra $3T_n(\beta)$. From Lemma 2.1 follows immediately that in the algebra $3HT_n$ the elements $\{u_{ij}\}$ satisfy the multiplicative (or quantum) Yang–Baxter relations

$$u_{ij}u_{ik}u_{jk} = u_{jk}u_{ik}u_{ij}, \quad \text{if } i < j < k. \tag{10}$$

Therefore one can define multiplicative analogues Θ_i, $1 \leq i \leq n$, of the Dunkl elements θ_i. Namely, to start with, we define elements $h_{ij} := h_{ij}(t) = 1 + tu_{ij}$, $i \neq j$. We consider $h_{ij}(t)$ as an element of the algebra $\widetilde{3HT_n} := 3HT_n \otimes \mathbb{Z}[[\beta, q_{ij}^{\pm 1}, t, x, y, \ldots]]$, where we assume that all parameters $\{\beta, q_{ij}, t, x, y, \ldots\}$ are central in the algebra $\widetilde{3HT_n}$.

Lemma 3.2.

(1a) $h_{ij}(x)h_{ij}(y) = h_{ij}(x+y+\beta xy) + q_{ij}xy$,

(1b) $h_{ij}(x)h_{ji}(y) = h_{ij}(x-y) + \beta y - q_{ij}xy$, if $i < j$.

It follows from (1b) that $h_{ij}(t)h_{ji}(t) = 1 + \beta t - t^2 q_{ij}$, if $i < j$, and therefore the elements $\{h_{ij}\}$ are invertible in the algebra $\widetilde{3HT_n}$.

(2) $h_{ij}(x)h_{jk}(y) = h_{jk}(y)h_{ik}(x) + h_{ik}(y)h_{ij}(x) - h_{ik}(x+y+\beta xy)$.

(3) *(Multiplicative Yang–Baxter relations)*

$$h_{ij}h_{ik}h_{jk} = h_{jk}h_{ik}h_{ij}, \quad \text{if } i < j < k.$$

(4) *Define multiplicative Dunkl elements (in the algebra $\widetilde{3HT_n}$) as follows:*

$$\Theta_j := \Theta_j(t) = \left(\prod_{a=j-1}^{1} h_{aj}^{-1} \right) \left(\prod_{a=n}^{j+1} h_{ja} \right), \quad 1 \leq j \leq n. \tag{11}$$

Then the multiplicative Dunkl elements pair-wise commute.

Clearly

$$\prod_{j=1}^{n} \Theta_j = 1, \quad \Theta_j = 1 + t\theta_j + t^2(\cdots), \quad \text{and} \quad \Theta_I \prod_{\substack{i \notin I, j \in I \\ i < j}} (1 + t\beta - t^2 q_{ij}) \in 3HT_n.$$

Here for a subset $I \subset [1, n]$ we use notation $\Theta_I = \prod_{a \in I} \Theta_a$.

Our main result of this Section is a description of relations among the multiplicative Dunkl elements.

Theorem 3.1. *In the algebra $3HT_n$ the following relations hold true*

$$\sum_{\substack{I \subset [1,n] \\ |I|=k}} \Theta_I \prod_{\substack{i \notin I, j \in J \\ i < j}} (1 + t\beta - t^2 q_{ij}) = \begin{bmatrix} n \\ k \end{bmatrix}_{1+t\beta}.$$

Here $\begin{bmatrix} n \\ k \end{bmatrix}_q$ denotes the q-Gaussian polynomial.

Corollary 3.1. *Assume that $q_{ij} \neq 0$ for all $1 \leq i < j \leq n$. Then the all elements $\{u_{ij}\}$ are invertible and $u_{ij}^{-1} = q_{ij}^{-1}(u_{ij} - \beta)$. Now define elements $\Phi_i \in \widehat{3HT_n}$ as follows:*

$$\Phi_i = \left\{ \prod_{a=i-1}^{1} u_{ai}^{-1} \right\} \left\{ \prod_{a=n}^{i+1} u_{ia} \right\}, \quad i = 1, \dots, n.$$

Then we have:

(1) *(Relationship among Θ_j and Φ_j)*

$$t^{n-2j+1} \Theta_j(t^{-1})|_{t=0} = (-1)^j \Phi_j.$$

(2) *The elements $\{\Phi_i, 1 \leq i \leq n,\}$ generate a commutative subalgebra in the algebra $\widehat{3HT_n}$.*

(3) *For each $k = 1, \dots, n$, the following relation in the algebra $3HT_n$ among the elements $\{\Phi_i\}$ holds:*

$$\sum_{\substack{I \subset [1,n] \\ |I|=k}} \prod_{\substack{i \notin I, j \in I \\ i < j}} (-q_{ij}) \Phi_I = \beta^{k(n-k)},$$

where $\Phi_I := \prod_{a \in I} \Phi_a$.

In fact the element Φ_i admits the following "reduced expression" which is useful for proofs and applications

$$\Phi_i = \left\{ \overrightarrow{\prod_{j \in I}} \left\{ \overrightarrow{\prod_{\substack{i \in I_+^c \\ i < j}}} u_{ij}^{-1} \right\} \right\} \left\{ \overrightarrow{\prod_{j \in I_+^c}} \left\{ \overrightarrow{\prod_{\substack{i \in I \\ i < j}}} u_{ij} \right\} \right\}. \tag{12}$$

Let us explain notations. For any (totally) ordered set $I = (i_1 < i_2 < \cdots < i_k)$ we denote by I_+ the set I with the opposite order, i.e., $I_+ = (i_k > i_{k-1} > \cdots > i_1)$; if $I \subset [1, n]$, then $I^c = [1, n] \backslash I$. For any (totally) ordered set I we denote by $\overrightarrow{\prod}_{i \in I}$ the ordered product according to the order of the set I.

Note that the total number of terms in the RHS of (12) is equal to $k(n-k) >$

Finally, from the "reduced expression" (12) for the element Φ_i one can see that

$$\prod_{\substack{i \notin I, j \in I \\ i<j}} (-q_{ij})\, \Phi_I = \left\{ \overrightarrow{\prod_{j \in I}} \left\{ \overrightarrow{\prod_{\substack{i \in I_+^c \\ i<j}}} (\beta - u_{ij}) \right\} \right\} \left\{ \overrightarrow{\prod_{j \in I_+^c}} \left\{ \overrightarrow{\prod_{\substack{i \in I \\ i<j}}} u_{ij} \right\} \right\}$$

$$:= \tilde{\Phi}_I \in 3HT_n\,.$$

Therefore the identity

$$\sum_{\substack{I \subset [1,n] \\ |I|=k}} \tilde{\Phi}_I = \beta^{k(n-k)}$$

is true in the algebra $3HT_n$ for arbitrary set of parameters $\{q_{ij}\}$.

Comments 3.2.

(I) In fact from our proof of Theorem 3.1 we can deduce more general statement, namely, consider integers m and k such that $1 \le k \le m \le n$. Then,

$$\sum_{\substack{I \subset [1,m] \\ |I|=k}} \Theta_I \prod_{\substack{i \in [1,m]\setminus I, j \in J \\ i<j}} (1 + t\beta - t^2 q_{ij}) = \begin{bmatrix} m \\ k \end{bmatrix}_{1+t\beta} + \sum_{\substack{A \subset [1,n], B \subset [1,n] \\ |A|=|B|=r}} u_{A,B}\,, \qquad (13)$$

where, by definition, for two sets $A = (i_1, \ldots, i_r)$ and $B = (j_1, \ldots, j_r)$ the symbol $u_{A,B}$ is equal to the (ordered) product $\prod_{a=1}^{r} u_{i_a, j_a}$. Moreover, the elements of the sets A and B have to satisfy the following conditions:

- for each $a = 1, \ldots, r$ one has $1 \le i_a \le m < j_a \le n$ and $k \le r \le k(n-k)$.

Even more, if $r = k$, then sets A and B have to satisfy the following additional conditions:

- $B = (j_1 \le j_2 \le \cdots \le j_k)$ and the elements of the set A are pair-wise distinct.

In the case $\beta = 0$ and $r = k$, i.e., in the case of additive (truncated) Dunkl elements, the above statement, also known as the quantum Pieri formula, has been stated as Conjecture in Ref. 7 and has been proved later in Ref. 25.

Corollary 3.2.[17]   *In the case when $\beta = 0$ and $q_{ij} = q_i \delta_{j-i,1}$, the algebra over $\mathbb{Z}[q_1, \ldots, q_{n-1}]$ generated by the multiplicative Dunkl elements $\{\Theta_i$ and $\Theta_i^{-1}, 1 \le i \le n\}$ is canonically isomorphic to the quantum K-theory of the complete flag variety Fl_n of type A_{n-1}.*

It is still an open problem to describe explicitly the set of monomials $\{u_{A,B}\}$ which appear in the RHS of (13) when $r > k$.

(II) (**Truncated Gaudin operators**) Let $\{p_{ij}1 \le i \ne j \le n\}$ be a set of mutually commuting parameters. We assume that parameters $\{p_{ij}\}$ are invertible and satisfy the Arnold relations

$$\frac{1}{p_{ik}} = \frac{1}{p_{ij}} + \frac{1}{p_{jk}}\,, \quad i < j, k\,.$$

For example one can take $p_{ij} = (z_i - z_j)^{-1}$, where $z = (z_1, \ldots, z_n) \in (\mathbb{C}\backslash 0)^n$.

Definition 3.3. Truncated (rational) Gaudin operator corresponding to the set of parameters $\{p_{ij}\}$, is defined to be:

$$G_i = \sum_{j \neq i} p_{ij}^{-1} s_{ij}, \quad 1 \leq i \leq n,$$

where s_{ij} denotes the exchange operator which switches variables x_i and x_j, and fixes parameters $\{p_{ij}\}$.

We consider the Gaudin operator G_i as an element of the group ring $\mathbb{Z}[\{p_{ij}^{\pm 1}\}][\mathbb{S}_n]$, call this element $G_i \in \mathbb{Z}[\{p_{ij}^{\pm 1}\}][\mathbb{S}_n]$, $i = 1, \ldots, n$, by Gaudin element and denoted it by $\theta_i^{(n)}$.

It is easy to see that the elements $u_{ij} := p_{ij}^{-1} s_{ij}$, $1 \leq i \neq j \leq n$, define a representation of the algebra $3HT_n$ with parameters $\beta = 0$ and $q_{ij} = u_{ij}^2 = p_{ij}^2$.

Therefore one can consider the (truncated) Gaudin elements as a special case of the (truncated) Dunkl elements. Now one can rewrite the relations among the Dunkl elements, as well as the quantum Pieri formula,[7,25] in terms of the Gaudin elements.

The key observation which allows to rewrite the quantum Pieri formula as a certain relation among the Gaudin elements is the following one: parameters $\{p_{ij}^{-1}\}$ satisfy the Plücker relations

$$\frac{1}{p_{ik} p_{jl}} = \frac{1}{p_{ij} p_{kl}} + \frac{1}{p_{il} p_{jk}}, \quad \text{if } i < j < k < l.$$

To describe relations among the Gaudin elements $\theta_i^{(n)}$, $i = 1, \ldots, n$, we need a bit of notation. Let $\{p_{ij}\}$ be a set of invertible parameters as before.

Define polynomials in the variables $\mathbf{h} = (h_1, \ldots, h_n)$

$$G_{m,k,r}(\mathbf{h}, \{p_{ij}\}) = \sum_{\substack{I \subset [1, n-1] \\ |I| = r}} \frac{1}{\prod_{i \in I} p_{in}} \sum_{\substack{J \subset [1, n] \\ |I| + m = |J| + k}} \binom{n - |I \cup J|}{n - m - |I|} \tilde{h}_J, \qquad (14)$$

where

$$\tilde{h}_J = \sum_{\substack{K \subset J, \ L \subset J, \\ |K| = |L|, \ K \cap L = \emptyset}} \prod_{j \in J \backslash (K \cup L)} h_j \prod_{k_a \in K, l_a \in L} p_{k_a, l_a}^2,$$

and summation runs over subsets $K = \{k_1, k_2 < \cdots < k_r\} \subset J$, and $L = \{l_a \in J, a = 1, \ldots, r\}$, such that $k_a < l_a, 1 \leq a \leq r$, and $l_1, \ldots, l_r$ are pairwise distinct.

Theorem 3.2. (Relations among the Gaudin elements,[14] cf. Ref. 24). *Under the assumption that elements $\{p_{ij}, 1 \leq i < j \leq n\}$ are invertible, mutually commute and satisfy the Arnold relations, one has:*

- $G_{m,k,r}(\theta_1^{(n)}, \ldots, \theta_n^{(n)}, \{p_{ij}\}) = 0, \quad \text{if } m > k,$ (15)

- $G_{0,0,k}(\theta_1^{(n)}, \ldots, \theta_n^{(n)}, \{p_{ij}\}) = e_k(d_2, \ldots, d_n)$, *where* $d_2, \ldots, d_n$ *denote the Jucys–Murphy elements in the group ring* $\mathbb{Z}[\mathbb{S}_n]$ *of the symmetric group* $\mathbb{S}_n$.

It is well-known that the elementary symmetric polynomials $e_k(d_2, \ldots, d_n) := C_k$, $k = 1, \ldots, n$, generate the center of the group ring $\mathbb{Z}[p_{ij}^{\pm 1}][\mathbb{S}_n]$, whereas the Gaudin elements $\{\theta_i^{(n)}, i = 1, \ldots, n\}$, generate a maximal commutative subalgebra $\mathcal{B}(p_{ij})$, the so-called <u>Bethe subalgebra</u>, in $\mathbb{Z}[p_{ij}^{\pm 1}][\mathbb{S}_n]$. It is well-known, see e.g., Ref. 24, that $\mathcal{B}(p_{ij}) = \oplus_{\lambda \vdash n} \mathcal{B}_\lambda(p_{ij})$, where $\mathcal{B}_\lambda(p_{ij})$ is the λ-isotypic component of $\mathcal{B}(p_{ij})$. On each λ-isotypic component the value of the central element C_k is the explicitly known constant $c_k(\lambda)$. It follows from Ref. 24 that the relations (15) together with relations

$$G_{0,0,k}(\theta_1^{(n)}, \ldots, \theta_n^{(n)}, \{p_{ij}\}) = c_k(\lambda)$$

are the defining relations for the algebra $\mathcal{B}_\lambda(p_{ij})$.

Let us remark that in the definition of the Gaudin elements we can use *any* set of mutually commuting, invertible elements $\{p_{ij}\}$ which satisfies the Arnold conditions. For example, we can take:

$$p_{ij} := \frac{q^{j-2}(1-q)}{1-q^{j-i}}, \quad 1 \le i < j \le n.$$

It is not difficult to see that in this case

$$\lim_{q \to 0} \frac{\theta_J^{(n)}}{p_{1j}} = -d_j = -\sum_{a=1}^{j-1} s_{aj},$$

where d_j denotes the Jucys–Murphy element in the group ring $\mathbb{Z}[\mathbb{S}_n]$ of the symmetric group $\mathbb{S}_n$. Basically from relations (15) one can deduce the relations among the Jucys–Murphy elements $d_2, \ldots, d_n$ after plugging in (15) the values $p_{ij} := (q^{j-2}(1-q)/1-q^{j-i})$ and passing to the limit $q \to 0$. However the real computations are rather involved.

Finally we note that the <u>multiplicative</u> Dunkl/Gaudin elements $\{\Theta_i, 1, \ldots, n\}$ also generate a maximal commutative subalgebra in the group ring $\mathbb{Z}[p_{ij}^{\pm 1}][\mathbb{S}_n]$. Some relations among the elements $\{\Theta_l\}$ follow from Theorem 2.1, but we don't know an analogue of relations (2.13) for the multiplicative Gaudin elements, but see Ref. 24.

(III) (Shifted Dunkl elements $\mathfrak{d}_i$ and $\mathfrak{D}_i$) As it was stated in Corollary 3.2, the *truncated* additive and multiplicative Dunkl elements in the algebra $3HT_n(0)$ generate over the ring of polynomials $\mathbb{Z}[q_1, \ldots, q_{n-1}]$ correspondingly the *quantumcohomology* and *quantumK − theory* rings of the full flag variety $\mathcal{F}l_n$. In order to describe the corresponding *equivariant* theories, we will introduce the *shifted* additive and multiplicative Dunkl elements. To start with we need at first to introduce an extension of the algebra $3HT_n(\beta)$.

Let $\{z_1, \ldots, z_n\}$ be a set of mutually commuting elements and $\{\beta, h, t, q_{ij} = q_{ji}, 1 \leq i, j \leq n\}$ be a set of parameters.

Definition 3.4. Define algebra $\overline{3TH_n}(\beta)$ to be the semi-direct product of the algebra $3TH_n(\beta)$ and the ring of polynomials $\mathbb{Z}[h, t][z_1, \ldots, z_n]$ with respect to the crossing relations

(1) $z_i u_{kl} = u_{kl} z_i$ if $i \notin \{k, l\}$,

(2) $z_i u_{ij} = u_{ij} z_j + \beta z_i + h$, $z_j u_{ij} = u_{ij} z_i - \beta z_i - h$, if $1 \leq i < j < k \leq n$.

Now we set as before $h_{ij} := h_{ij}(t) = 1 + t u_{ij}$.

Definition 3.5.

- Define shifted additive Dunkl elements to be:

$$\mathfrak{d}_i = z_i - \sum_{i<j} u_{ij} + \sum_{i<j} u_{ji}.$$

- Define shifted multiplicative Dunkl elements to be:

$$\mathfrak{D}_i = \left(\prod_{a=i-1}^{1} h_{ai}^{-1} \right) (1 + z_i) \left(\prod_{a=n}^{i+1} h_{ia} \right).$$

Lemma 3.3.

$$[\mathfrak{d}_i, \mathfrak{d}_j] = 0, \quad [\mathfrak{D}_i, \mathfrak{D}_j] = 0 \quad \text{for all } i, j.$$

Now we stated an analogue of Theorem 3.1 for shifted multiplicative Dunkl elements. As a preliminary, for any subset $I \subset [1, n]$ let us set $\mathfrak{D}_I = \prod_{a \in I} \mathfrak{D}_a$. It is clear that:

$$\mathfrak{D}_I \prod_{\substack{i \notin I, j \in I \\ i<j}} (1 + t\beta - t^2 q_{ij}) \in \overline{3HT_n}(\beta).$$

Theorem 3.3. *In the algebra $\overline{3HT_n}(\beta)$ the following relations hold true*

$$\sum_{\substack{I \subset [1,n] \\ |I|=k}} \mathfrak{D}_I \prod_{\substack{i \notin I, j \in J \\ i<j}} (1 + t\beta - t^2 q_{ij}) = \begin{bmatrix} n \\ k \end{bmatrix}_{1+t\beta}$$

$$+ \sum_{\substack{I \subset [1,n] \\ I = \{i_1, \ldots, i_k\}}} \prod_{a=1}^{k} \left[z_a (1 + \beta t)^{n-k} + h \frac{(1 + \beta t)^{n-k} - (1 + \beta t)^{i_a - a}}{\beta} \right].$$

In particular, if $\beta = 0$, we will have

Corollary 3.3. *In the algebra $\overline{3HT_n}(0)$ the following relations hold*

$$\sum_{\substack{I \subset [1,n] \\ |I|=k}} \mathfrak{D}_I \prod_{\substack{i \notin I, j \in J \\ i<j}} (1 - t^2 q_{ij}) = \binom{n}{k} + \sum_{\substack{I \subset [1,n] \\ I = \{i_1, \ldots, i_k\}}} \prod_{a=1}^{k} \prod_{a=1}^{n} (z_a + th(n - k - i_a + a)).$$

One of the main steps in our proof of Theorem 3.3 is the following explicit formula for the elements $\mathfrak{D}_I$.

Lemma 3.4. *One has*

$$\tilde{\mathfrak{D}}_I := \mathfrak{D}_I(1 + t\beta - t^2 q_{ij}) = \prod_{b \in I}^{\nearrow} \left(\prod_{\substack{a \notin I \\ a < b}}^{\searrow} h_{ba} \right) \prod_{a \in I}^{\nearrow} \left((1 + z_a) \prod_{\substack{b \notin I \\ a < b}}^{\searrow} h_{ab} \right).$$

Note that if $a < b$, then $h_{ba} = 1 + \beta t - u_{ab}$. Here we have used the symbol

$$\prod_{b \in I}^{\nearrow} \left(\prod_{\substack{a \notin I \\ a < b}}^{\searrow} h_{ba} \right)$$

to denote the following product. At first, for a given element $b \in I$ let us define the set $I(b) := \{a \in [1, n] \backslash I, a < b\} := (a_1^{(b)} < \cdots < a_p^{(b)})$ for some p (depending on b). If $I = (b_1 < b_2 \cdots < b_k)$, then we set:

$$\prod_{b \in I}^{\nearrow} \left(\prod_{\substack{a \notin I \\ a < b}}^{\searrow} h_{ba} \right) = \prod_{j=1}^{k} (u_{b_j, a_s} u_{b_j, a_{s-1}} \cdots u_{b_j, a_1}).$$

For example, let us take $n = 6$ and $I = (1, 3, 5)$. Then,

$$\tilde{\mathfrak{D}}_I = h_{32} h_{54} h_{52} (1 + z_1) h_{16} h_{14} h_{12} (1 + z_3) h_{36} h_{34} (1 + z_5) h_{56}.$$

4. Combinatorics of Associative Quasi-Classical Yang–Baxter Algebras

Let β be a parameter.

Definition 4.1.[14] The associative quasi-classical Yang–Baxter algebra of weight β, denoted by $\widehat{ACYB}_n(\beta)$, is an associative algebra, over the ring of polynomials $\mathbb{Z}[\beta]$, generated by the set of elements $\{x_{ij}, 1 \le i \ne j \le n\}$, subject to the set of relations

(a) $x_{ij} x_{kl} = x_{kl} x_{ij}$, if $\{i, j\} \cap \{k, l\} = \emptyset$,
(b) $x_{ij} x_{jk} = x_{ik} x_{ij} + x_{jk} x_{ik} + \beta x_{ik}$, if $1 \le 1 < i < j \le n$,
(c) $x_{ij} + x_{ji} = \beta$.

Comments 4.1. The algebra $3T_n(\beta)$, see Definition 3.1, is the quotient of the algebra $\widehat{ACYB}_n(-\beta)$, by the "dual relations"

$$x_{jk} x_{ij} - x_{ij} x_{ik} - x_{ik} x_{jk} + \beta x_{ik} = 0, \quad i < j < k.$$

The (truncated) Dunkl elements $\theta_i = \sum_{j \ne i} x_{ij}$, $i = 1, \ldots, n$, do not commute in the algebra $\widehat{ACYB}_n(\beta)$. However a certain version of noncommutative elementary polynomial of degree $k \ge 1$, still is equal to zero after the substitution of Dunkl elements instead of variables.[14] We state here the corresponding result only "in classical case", i.e., if $\beta = 0$ and $q_{ij} = 0$ for all i, j.

Lemma 4.1.[14] *Define noncommutative elementary polynomial $L_k(x_1, \ldots, x_n)$ as follows*

$$L_k(x_1, \ldots, x_n) = \sum_{I=(i_1 < i_2 < \cdots < i_k) \subset [1,n]} x_{i_1} x_{i_2} \cdots x_{i_k} .$$

Then $L_k(\theta_1, \theta_2, \ldots, \theta_n) = 0$.

Moreover, if $1 \le k \le m \le n$, then one can show that the value of the noncommutative polynomial $L_k(\theta_1, \ldots, \theta_m)$ in the algebra $\widehat{ACYB}_n(\beta)$ is given by the Pieri formula.[7,25]

4.1. *Combinatorics of Coxeter element*

Consider the "Coxeter element" $w \in \widehat{ACYB}_n(\beta)$ which is equal to the ordered product of "simple generators": $w := \prod_{a=1}^{n-1} x_{a,a+1}$. Let us bring the element w to the reduced form in the algebra $\widehat{ACYB}_n(\beta)$, that is, let us consecutively apply the defining relations (a) and (b) to the element w in any order until unable to do so. Denote the resulting (noncommutative) polynomial by $P(x_{ij}; \beta)$. In principal, the polynomial itself can depend on the order in which the relations (a) and (b) are applied.

Proposition 4.1.[23] Cf. Ref. 28, 8.C5, (c))

(**1**) *Apart from applying the relation (a) (commutativity), the polynomial $P(x_{ij}; \beta)$ does not depend on the order in which relations (a) and (b) have been applied, and can be written in a unique way as a linear combination:*

$$P_n(x_{ij}; \beta) = \sum_{s=1}^{n-1} \beta^{n-s-1} \sum_{\{i_a\}} \prod_{a=1}^{s} x_{i_a, j_a} ,$$

where the second summation runs over all sequences of integers $\{i_a\}_{a=1}^{s}$ such that $n - 1 \ge i_1 \ge i_2 \ge \cdots \ge i_s = 1$, and $i_a \le n - a$ for $a = 1, \ldots, s - 1$; moreover, the corresponding sequence $\{j_a\}_{a=1}^{n-1}$ can be defined uniquely by that $\{i_a\}_{a=1}^{n-1}$.

- *It is clear that the polynomial $P(x_{ij}; \beta)$ also can be written in a unique way as a linear combination of monomials $\prod_{a=1}^{s} x_{i_a, j_a}$ such that $j_1 \ge j_2 \ldots \ge j_s$.*

(**2**) *Denote by $T_n(k, r)$ the number of degree k monomials in the polynomial $P(x_{ij}; \beta)$ which contain exactly r factors of the form $x_{*,n}$. (Note that $1 \le r \le k \le n - 1$). Then*

$$T_n(k, r) = \frac{r}{k} \binom{n + k - r - 2}{n - 2} \binom{n - 2}{k - 1} .$$

In particular, $T_n(k, k) = \binom{n-2}{k-1}$, and $T_n(k, 1) = T(n - 2, k - 1)$, where $T(n, k) := (1/k + 1) \binom{n+k}{k} \binom{n}{k}$ is equal to the number of Schröder paths (i.e., consisting of steps

$U = (1,1)$, $D = (1,-1)$, $H = (2,0)$ *and never going below the x-axis) from* $(0,0)$
to $(2n, 0)$, *having* kU's, *see Ref. 27, A088617.*

Moreover, $T_n(n-1, r) = Tab(n-2, r-1)$, where $Tab(n, k) := (k+1)/(n+1)\binom{2n-k}{n}$ *is equal to the number of standard Young tableaux of the shape* $(n, n-k)$, *see Ref. 27, A009766.*

(3) *After the specialization* $x_{ij} \to 1$ *the polynomial* $P(x_{ij})$ *is transformed to the polynomial*

$$P_n(\beta) := \sum_{k=0}^{n-1} N(n,k)(1+\beta)^k ,$$

where $N(n,k) := (1/n)\binom{n}{k}\binom{n}{k+1}$, $k = 0, \ldots, n-1$, *stand for the Narayana numbers. Furthermore,* $P_n(\beta) = \sum_{d=0}^{n-1} s_n(d)\,\beta^d$, *where* $s_n(d) = 1/(n+1)\binom{2n-d}{n}\binom{n-1}{d}$ *is the number of ways to draw* $n - 1 - d$ *diagonals in a convex* $(n+2)$-*gon, such that no two diagonals intersect their interior.*

Therefore, the number of (nonzero) terms in the polynomial $P(x_{ij}; \beta)$ *is equal to the nth little Schröder number* $s_n := \sum_{d=0}^{n-1} s_n(d)$, *also known as the nth super-Catalan number, see e.g., Ref. 27, A001003.*

(4) *After the specialization* $x_{1j} \to t$, $1 \le j \le n$, *and that* $x_{ij} \to 1$, *if* $2 \le i < j \le n$, *the polynomial* $P(x_{ij}; \beta)$ *is transformed to the polynomial*

$$P_n(\beta, t) = t \sum_{k=1}^{n} (1+\beta)^{n-k} \sum_{\pi} t^{p(\pi)} ,$$

where the second summation runs over the set of Dick paths π *of length* $2n$ *with exactly k picks (UD-steps), and* $p(\pi)$ *denotes the number of valleys (DU-steps) that touch upon the line* $x = 0$.

(5) *The polynomial* $P(x_{ij}; \beta)$ *is invariant under the action of anti-involution* $\phi \circ \tau$, *see Sec. 5.1.1*[14] *for definitions of* ϕ *and* τ.

(6) *Follow Refs. 28,* **6.C8**, *(c), consider the specialization*

$$x_{ij} \to t_i, \quad 1 \le i < j \le n,$$

and define $P_n(t_1, \ldots, t_{n-1}; \beta) = P_n(x_{ij} = t_i; \beta)$.

One can show, ibid, that

$$P_n(t_1, \ldots, t_{n-1}; \beta) = \sum \beta^{n-k} \; t_{i_1} \cdots t_{i_k} , \tag{16}$$

where the sum runs over all pairs $\{(a_1, \ldots, a_k), (i_1, \ldots, i_k) \in \mathbb{Z}_{\ge 1} \times \mathbb{Z}_{\ge 1}\}$ *such that* $1 \le a_1 < a_2 < \cdots < a_k$, $1 \le i_1 \le i_2 \cdots \le i_k \le n$ *and* $i_j \le a_j$ *for all* j.

To continue, let $\pi \in \mathbb{S}_n$ *be the permutation* $\pi = \left(\begin{smallmatrix} 1 & 2 & 3 & \cdots & n \\ 1 & n & n-1 & \cdots & 2 \end{smallmatrix}\right)$. *Then*

$$P_n(t_1, \ldots, t_{n-1}; \beta) = \prod_{i=1}^{n-1} t_i^{n-i} \mathfrak{G}_\pi^{(\beta)}(t_1^{-1}, \ldots, t_{n-1}^{-1}),$$

where $\mathfrak{G}_w^{(\beta)}(x_1, \ldots, x_{n-1})$ *denotes the* β-*Grothendieck polynomial corresponding to a permutation* $w \in \mathbb{S}_n$.[8]

In particular,

$$\mathfrak{G}_\pi^{(\beta)}(x_1 = 1, \ldots, x_{n-1} = 1) = \sum_{k=0}^{n-1} N(n,k)(1+\beta)^k \,,$$

where $N(n,k)$ denotes the Narayana numbers, see item **(3)** *of Proposition 4.1.*

- Note that if $\beta = 0$, then one has $\mathfrak{G}_w^{(\beta=0)}(x_1, \ldots, x_{n-1}) = \mathfrak{S}_w(x_1, \ldots, x_{n-1})$, that is the β-Grothendieck polynomial at $\beta = 0$, is equal to the Schubert polynomial corresponding to the same permutation w. Therefore, if $\pi = \left(\begin{smallmatrix} 1 & 2 & 3 & \ldots & n \\ 1 & n & n-1 & \ldots & 2 \end{smallmatrix}\right)$, then

$$\mathfrak{S}_\pi(x_1 = 1, \ldots, t_{n-1} = 1) = C_{n-1} \,, \tag{17}$$

where C_m denotes the mth Catalan number. Using the formula (16) it is not difficult to check the following formula for the principal specialization of the Schubert polynomial $\mathfrak{S}_\pi$

$$\mathfrak{S}_\pi(1, q, \ldots, q^{n-1}) = q^{\binom{n-1}{3}} C_{n-1}(q) \,, \tag{18}$$

where $C_m(q)$ denotes the Carlitz–Riordan q-analogue of the Catalan numbers.[29] The formula (17) has been proved in Ref. 10 using the observation that π is a *vexillary* permutation, see Ref. 21 for the definition of the latter. A combinatorial/bijective proof of the formula (18) is due to Woo.[33]

Comments 4.2. The Grothendieck polynomials defined by Lascoux and Schützenberger,[20] correspond to the case $\beta = -1$. In this case $P_n(-1) = 1$, if $n \geq 0$, and therefore the specialization $\mathfrak{G}_w^{(-1)}(x_1 = 1, \ldots, x_{n-1} = 1) = 1$ for all $w \in \mathbb{S}_n$.

4.2. *Grothendieck and q-Schröderpolynomials*

4.2.1. *Schröder paths and polynomials*

Definition 4.2. A Schröder path of the length n is an over diagonal path from $(0,0)$ to (n,n) with steps $(1,0)$, $(0,1)$ and steps $D = (1,1)$ <u>without</u> steps of type D on the diagonal $x = y$.

If p is a Schröder path, we denote by $d(p)$ the number of the diagonal steps of the path p, and by $a(p)$ the number of unit squares located between the path p and the diagonal $x = y$. For each (unit) diagonal step D of a path p we denote by $i(D)$ the x-coordinate of the column which contains the diagonal step D. Finally, define the index $i(p)$ of a path p as the some of the numbers $i(D)$ for all diagonal steps of the path p.

Definition 4.3. Define q-Schröder polynomial $S_n(q; \beta)$ as follows:

$$S_n(q; \beta) = \sum_p q^{a(p)+i(p)} \beta^{d(p)} \,, \tag{19}$$

where the sum runs over the set of all Schröder paths of length n.

Example 4.1.

$$S_1(q;\beta) = 1, \quad S_2(q;\beta) = 1 + q + \beta q,$$

$$S_3(q;\beta) = 1 + 2q + q^2 + q^3 + \beta(q + 2q^2 + 2q^3) + \beta^2 q^3,$$

$$S_4(q;\beta) = 1 + 3q + 3q^2 + 3q^3 + 2q^4 + q^5 + q^6 + \beta(q + 3q^2 + 5q^3 + 6q^4 + 3q^5 + 3q^6)$$

$$+ \beta^2(q^3 + 2q^4 + 3q^5 + 3q^6) + \beta^3 q^6.$$

Comments 4.3. The q-Schröder polynomials defined by the formula (19) are _different_ from the q-analogue of Schröder polynomials which has been considered in Ref. 3. It seems that there are no simple connections between the both.

Proposition 4.2. (*Recurrence relations for q-Schröder polynomials*) *The Schröder polynomials satisfy the following relations*

$$S_{n+1}(q;\beta) = (1 + q^n + \beta\, q^n)S_n(q;\beta)$$

$$+ \sum_{k=1}^{k=n-1} (q^k + \beta\, q^{n-k})S_k(q; q^{n-k}\beta)S_{n-k}(q;\beta) \tag{20}$$

and the initial condition $S_1(q;\beta) = 1$.

Note that $P_n(\beta) = S_n(1;\beta)$ and in particular, the polynomials $P_n(\beta)$ satisfy the following recurrence relations

$$P_{n+1}(\beta) = (2 + \beta)P_n(\beta) + (1 + \beta)\sum_{k=1}^{n-1} P_k(\beta)P_{n-k}(\beta). \tag{21}$$

Theorem 4.1. (*Evaluation of the Schröder–Hankel Determinant*) *Consider permutation*

$$\pi_k^{(n)} = \begin{pmatrix} 1 & 2 & \cdots & k & k+1 & k+2 & \cdots & n \\ 1 & 2 & \cdots & k & n & n-1 & \cdots & k+1 \end{pmatrix}.$$

Let as before

$$P_n(\beta) = \sum_{j=0}^{n-1} N(n,j)(1+\beta)^j, \quad n \geq 1, \tag{22}$$

denotes the Narayana–Schröder polynomials. _Then_

$$(1+\beta)^{\binom{k}{2}}\mathfrak{S}_{\pi_k^{(n)}}^{(\beta)}(x_1 = 1, \ldots, x_{n-k} = 1) = \mathrm{Det}|P_{n+k-i-j}(\beta)|_{1\leq i,j\leq k}. \tag{23}$$

Proof is based on an observation that the permutation $\pi_k^{(n)}$ is a *vexillary* one and the recurrence relations (21).

Comments 4.4.

(1) In the case $\beta = 0$, i.e., in the case of <u>Schubert polynomials</u> Theorem 4.1 has been proved in Ref. 10.

(2) In the cases when $\beta = 1$ and $0 \le n - k \le 2$, the value of the determinant in the RHS(3.22) is known,[3] or Ichikawa talk *Hankel determinants of Catalan, Motzkin and Schröder numbers and its q-analogues.*[a] One can check that in the all cases mentioned above, the formula (22) gives the same results.

(3) **Grothendieck and Narayana polynomials**

It follows from the expression (22) for the Narayana–Schröder polynomials that $P_n(\beta - 1) = \mathfrak{N}_n(\beta)$, where

$$\mathfrak{N}_n(\beta) := \sum_{j=0}^{n-1} \frac{1}{n}\binom{n}{j}\binom{n}{j+1}\beta^j,$$

denotes the nth Narayana polynomial. Therefore, $P_n(\beta-1) = \mathfrak{N}_n(\beta)$ is a symmetric polynomial in β with nonnegative integer coefficients. Moreover, the value of the polynomial $P_n(\beta - 1)$ at $\beta = 1$ is equal to the nth Catalan number $C_n := 1/(n+1)\binom{2n}{n}$.

It is well-known,[31] that the Narayana polynomial $\mathfrak{N}_n(\beta)$ is equal to the generating function of the statistics $\pi(\mathfrak{p}) = (number\ of\ \underline{peaks}\ of\ a\ Dick\ path\ \mathfrak{p}) - 1$ on the set $Dick_n$ of Dick paths of the length $2n$

$$\mathfrak{N}_n(\beta) = \sum_{\mathfrak{p}} \beta^{\pi(\mathfrak{p})}.$$

Moreover, using the Lindström–Gessel–Viennot lemma[b] one can see that:

$$\mathrm{DET}|\mathfrak{N}_{n+k-i-j}(\beta)|_{1\le i,j\le k} = \beta^{\binom{k}{2}} \sum_{(\mathfrak{p}_1,\dots,\mathfrak{p}_k)} \beta^{\pi(\mathfrak{p}_1)+\cdots+\pi(\mathfrak{p}_k)}, \qquad (24)$$

where the sum runs over k-tuple of noncrossing Dick paths $(\mathfrak{p}_1,\dots,\mathfrak{p}_k)$ such that the path $\mathfrak{p}_i$ starts from the point $(i-1,0)$ and has length $2(n-i+1)$, $i = 1,\dots,k$.

We <u>denote</u> the sum in the RHS(24) by $\mathfrak{N}_n^{(k)}(\beta)$. Note that $\mathfrak{N}_{k-1}^{(k)}(\beta) = 1$ for all $k \ge 2$.

Thus, $\mathfrak{N}_n^{(k)}(\beta)$ is a symmetric polynomial in β with nonnegative integer coefficients, and

$$\mathfrak{N}_n^{(k)}(\beta = 1) = C_n^{(k)} = \prod_{1\le i<j\le n-k+2} \frac{2k+i+j-1}{i+j-1}.$$

As a corollary we obtain the following statement:

Proposition 4.3. *Let* $n \ge k$, *then,*

$$\mathfrak{G}_{\pi_k^{(n)}}^{(\beta-1)}(x_1 = 1,\dots,x_n = 1) = \mathfrak{N}_n^{(k)}(\beta).$$

---

[a]http:/denjoy.ms.u-tokyo.ac.jp.
[b]http://en.wikipedia.org/wiki/Lindström–Gessel–Viennot lemma

Summarizing, the specialization $\mathfrak{G}^{(\beta-1)}_{\pi^{(n)}_k}(x_1 = 1, \ldots, x_n = 1)$ is a symmetric polynomial in β with nonnegative integer coefficients, and coincides with the generating function of the statistics $\sum_{i=1}^{k} \pi(\mathfrak{p}_i)$ on the set k-$Dick_n$ of k-tuple of noncrossing Dick paths $(\mathfrak{p}_1, \ldots, \mathfrak{p}_k)$.

Example 4.2. Take $n = 5$, $k = 1$. Then $\pi^{(5)}_1 = (15432)$ and one has:

$$\mathfrak{G}^{(\beta)}_{\pi^{(5)}_1}(1, q, q^2, q^3) = q^4(1, 3, 3, 3, 2, 1, 1) + q^5(1, 3, 5, 6, 3, 3)\beta$$

$$+ q^7(1, 2, 3, 3)\beta^2 + q^{10}\beta^3.$$

It is easy to compute the Carlitz–Riordan q-analogue of the Catalan number C_5, namely,

$$C_5(q) = (1, 3, 3, 3, 2, 1, 1).$$

(4) **Grothendieck polynomials** $\mathfrak{G}^{(\beta)}_{\pi^{(n)}_k}(x_1, \ldots, x_n)$ **and k-dissections**

Let $k \in \mathbb{N}$ and $n \geq k-1$, be a integer, define a k-*dissection* of a convex $(n+k+1)$-gon to be a collection $\mathcal{E}$ of diagonals in $(n+k+1)$-gon not containing $(k+1)$-subset of pairwise crossing diagonals and such that at least $2(k-1)$ diagonals are coming from each vertex of the $(n+k+1)$-gon in question. One can show that the number of diagonals in any k-dissection $\mathcal{E}$ of a convex $(n+k+1)$-gon contains at least $(n+k+1)(k-1)$ and at most $n(2k-1)-1$ diagonals. We define the *index* of a k-dissection $\mathcal{E}$ to be $i(\mathcal{E}) = \setminus(\mathcal{E}\| - \infty) - \infty - \#|\mathcal{E}|$. Denote by:

$$\mathcal{T}^{(k)}_n(\beta) = \sum_{\mathcal{E}} \beta^{i(\mathcal{E})}$$

the generating function for the number of k-dissections with a fixed index, where the above sum runs over the set of all k-dissections of a convex $(n+k+1)$-gon.

Theorem 4.2.

$$\mathfrak{G}^{(\beta)}_{\pi^{(n)}_k}(x_1 = 1, \ldots, x_n = 1) = \mathcal{T}^{(k)}_n(\beta).$$

A k-dissection of a convex $(n+k+1)$-gon with the maximal number of diagonals (which is equal to $n(2k-1)-1$), is called k-*triangulation*. It is well-known that the number of k-triangulations of a convex $(n+k+1)$-gon is equal to the Catalan-Hankel number $C^{(k)}_{n-1}$. Explicit bijection between the set of k-triangulations of a convex $(n+k+1)$-gon and the set of k-tuple of non-crossing Dick paths $(\gamma_1, \ldots, \gamma_k)$ such that the Dick path γ_i connects points $(i-1, 0)$ and $(2n-i-1, 0)$, has been constructed in Refs. 26, and 30.

Problems 4.1.

(1) Define a bijection between monomials of the form $\prod_{a=1}^{s} x_{i_a, j_a}$ involved in the polynomial $P(x_{ij}; \beta)$, and dissections of a convex $(n+2)$-gon by s diagonals, such that no two diagonals intersect their interior.

(2) Describe permutations $w \in \mathbb{S}_n$ such that the Grothendieck polynomial $\mathfrak{G}_w(t_1, \ldots, t_n)$ is equal to the "reduced polynomial" for a some monomial in the associative qasi-classical Yang–Baxter algebra $\widehat{ACYB}_n(\beta)$.

(3) Study "reduced polynomial" corresponding to the monomials

$$u_{12}u_{23} \cdots u_{n-1,n}u_{n-2,n-1} \cdots u_{23}u_{12}, \quad (u_{12}u_{23} \cdots u_{n-1,n})^k$$

in the algebra $\widehat{ACYB}_n(\beta)^{ab}$.

(4) Construct a bijection between the set of k-dissections of a convex $(n + k + 1)$-gon and "pipe dreams" corresponding to the Grothendieck polynomial $\mathfrak{G}_{\pi_k^{(n)}}^{(\beta)}(x_1, \ldots, x_n)$. As for a definition of "pipe greams" for Grothendieck polyno-mials.[8,19]

(5) Is it true that polynomial $\mathfrak{G}_w^{(\beta-)}(x_1 = 1, \ldots, x_n = 1) \in \mathbb{N}[\beta]$ for any permu-tation $w \in \mathbb{S}_n$?

4.2.2. *Principal specialization of Grothendieck polynomials and q-Schröder polynomials*

Let $\pi_k^{(n)} = 1^k \times w_0^{(n-k)} \in \mathbb{S}_n$ be the vexillary permutation as before, see Theorem 4.1. Recall that:

$$\pi_k^{(n)} = \begin{pmatrix} 1 & 2 & \ldots & k & k+1 & k+2 & \ldots & n \\ 1 & 2 & \ldots & k & n & n-1 & \ldots & k+1 \end{pmatrix}.$$

(A) Principal specialization of the Schubert polynomial $\mathfrak{G}_{\pi_k^{(n)}}$

Note that $\pi_k^{(n)}$ is a vexillary permutation of the staircase shape $\lambda = (n - k - 1, \ldots, 2, 1)$ and has the staircase flag $\phi = (k+1, k+2, \ldots, n-1)$. It is known,[21,32] that for a vexillary permutation $w \in \mathbb{S}_n$ of the shape λ and flag $\phi = (\phi_1, \ldots, \phi_r)$, $r = \ell(\lambda)$, the corresponding Schubert polynomial $\mathfrak{G}_w(X_n)$ is equal to the multi-Schur polynomial $s_\lambda(X_\phi)$, where X_ϕ denotes the flagged set of variables, namely, $X_\phi = (X_{\phi_1}, \ldots, X_{\phi_r})$ and $X_m = (x_1, \ldots, x_m)$. Therefore we can write the folliing determinantal formula for the principal specialization of the Schubert polynomial corresponding to the vexillary permutation $\pi_k^{(n)}$

$$\mathfrak{G}_{\pi_k^{(n)}}(1, q, q^2, \ldots) = \mathrm{DET}\left(\begin{bmatrix} n-i+j-1 \\ k+i-1 \end{bmatrix}_q \right)_{1 \le i,j \le n-k},$$

where $\begin{bmatrix} n \\ k \end{bmatrix}_q$ denotes the q-binomial coefficient.

Let us observe that the Carlitz–Riordan q-analogue $C_n(q)$ of the Catalan number C_n is equal to the value of the q-Schröder polynomial at $\beta = 0$, namely, $C_n(q) = S_n(q, 0)$.

Lemma 4.2.

$$(1) \quad \mathrm{DET}\left(\begin{bmatrix} n-i+j-1 \\ k+i-1 \end{bmatrix}_q \right)_{1 \le i,j \le n-k} = q^{\binom{n-k}{2}} C_{n(k)}(q),$$

(2) $\text{DET}(C_{n+k-i-j}(q))_{1\leq i,j\leq k} = q^{k(k-1)(4k-1)/6} C_{n^{(k)}}(q)$.

(B) Principal specialization of the Grothendieck polynomial $\mathfrak{G}^{(\beta)}_{\pi_k^{(n)}}$

Theorem 4.3.

$$q^{\binom{n-k+1}{3}-(k-1)\binom{n-k}{2}}\text{DET}|S_{n+k-i-j}(q;\beta^{i-1})|_{1\leq i,j\leq k}$$

$$= q^{k(k-1)(4k+1)/6}\prod_{a=1}^{k-1}(1+q^{a-1})\mathfrak{G}_{\pi_k^{(n)}}(1,q,q^2,\dots).$$

Corollary 4.1. (1) *If* $k = n - 1$, *then,*

$$\text{DET}|S_{2n-1-i-j}(q;q^{i-1}\beta)|_{1\leq i,j\leq n-1} = q^{(n-1)(n-2)(4n-3)/6}\prod_{a=1}^{n-2}(1+q^{a-1}\beta)^{n-a-1}.$$

(2) *If* $k = n - 2$, *then,*

$$q^{n-2}\text{DET}|S_{2n-2-i-j}(q;q^{i-1}\beta)|_{1\leq i,j\leq n-2}$$

$$= q^{(n-2)(n-3)(4n-7)/6}\prod_{a=1}^{n-3}(1+q^{a-1}\beta)^{n-a-2}\left\{\frac{(1+\beta)^{n-1}-1}{\beta}\right\}.$$

• Generalization

Let $\mathbf{n} = (n_1, \dots, n_p) \in \mathbb{N}^p$ be a composition of n so that $n = n_1 + \cdots + n_p$. We set $n^{(j)} = n_1 + \cdots + n_j$, $j = 1, \dots, p$, $n^{(0)} = 0$.

Now consider the permutation $w^{(\mathbf{n})} = w_0^{(n_1)} \times w_0^{(n_2)} \times \cdots \times w_0^{(n_p)} \in \mathbb{S}_n$, where $w_0^{(m)} \in \mathbb{S}_m$ denotes the longest permutation in the symmetric group $\mathbb{S}_m$. In other words,

$$w^{(\mathbf{n})} = \begin{pmatrix} 1 & 2 & \cdots & n_1 & n^{(2)} & \cdots & n_1+1 & \cdots & n^{(p-1)} & \cdots & n \\ n_1 & n_1-1 & \cdots & 1 & n_1+1 & \cdots & n^{(2)} & \cdots & n & \cdots & n^{(p-1)}+1 \end{pmatrix}.$$

For the permutation $w^{(\mathbf{n})}$ defined above, one has the following factorization formula for the Grothendieck polynomial corresponding to $w^{(\mathbf{n})}$,[21]

$$\mathfrak{G}^{(\beta)}_{w^{(\mathbf{n})}} = \mathfrak{G}^{(\beta)}_{w_0^{(n_1)}} \times \mathfrak{G}^{(\beta)}_{1^{n_1}\times w_0^{(n_2)}} \times \mathfrak{G}^{(\beta)}_{1^{n_1+n_2}\times w_0^{(n_3)}} \times \cdots \times \mathfrak{G}^{(\beta)}_{1^{n_1+\cdots n_{p-1}}\times w_0^{(n_p)}}.$$

In particular, if,

$$w = w_0^{(n_1)} \times w_0^{(n_2)} \times \cdots \times w_0^{(n_p)} \in \mathbb{S}_n,$$

then the principal specialization $\mathfrak{G}^{(\beta)}_w$ of the Grothendieck polynomial corresponding to the permutation w, is the product of q-Schröder–Hankel polynomials. Finally, we observe that from discussions in Sec. 3.4, **Grothendieck and Narayana polynomials**, one can deduce that:

$$\mathfrak{G}^{(\beta-1)}_w(x_1 = 1, \dots, x_n = 1) = \prod_{j=1}^{p-1}\mathfrak{N}^{(n^{(j+1)})}_{n^{(j)}}(\beta).$$

In particular, the polynomial $\mathfrak{G}_w^{(\beta-1)}(x_1, \ldots, x_n)$ is a symmetric polynomial in β with nonnegative integer coefficients.

Example 4.3. Let us take (nonvexillary) permutation $w = 2143 = s_1 s_3$. One can check that $\mathfrak{G}_w^{(\beta)}(1,1,1,1) = 3 + 3\beta + \beta^2 = 1 + (\beta+1) + (\beta+1)^2$, and $\mathfrak{N}_4(\beta) = (1,6,6,1)$, $\mathfrak{N}_3(\beta) = (1,3,1)$, $\mathfrak{N}_2(\beta) = (1,1)$. It is easy to see that:

$$\beta \mathfrak{G}_w^{(\beta)}(1,1,1,1) = \text{DET} \begin{vmatrix} \mathfrak{N}_4(\beta) & \mathfrak{N}_3(\beta) \\ \mathfrak{N}_3(\beta) & \mathfrak{N}_2(\beta) \end{vmatrix}.$$

On the other hand, $\text{DET}\begin{vmatrix} P_4(\beta) & P_3(\beta) \\ P_3(\beta) & P_2(\beta) \end{vmatrix} = (3,6,4,1) = \underline{(3 + 3\beta + \beta^2)(1 + \beta)}$. It is more involved to check that:

$$q^5(1+\beta)\mathfrak{G}_w^{(\beta)}(1, q, q^2, q^3) = \text{DET} \begin{vmatrix} S_4(q; \beta) & S_3(q; \beta) \\ S_3(q; q\beta) & S_2(q; q\beta) \end{vmatrix}.$$

Comments 4.5. One can compute the Grothendieck polynomials for yet another interesting family of permutations. namely, permutations $\sigma_k^{(n)} =$

$$\sigma_k^{(n)} = \begin{pmatrix} 1 & 2 & \ldots & k-1 & k & k+1 & k+2 & \ldots & n+k \\ 1 & 2 & \ldots & k-1 & n+k & k & k+1 \ldots & n+k-1 \end{pmatrix}$$

$$= s_k s_{k+1} \ldots s_{n+k-1} \in \mathbb{S}_{n+k}.$$

<u>Then</u>

$$\mathfrak{G}_{\sigma_k}^{(n)}(x_1, \ldots, x_{n+k}) = \sum_{j=0}^{k-1} \binom{n+j-1}{j} e_{n+j}(x_1, \ldots, x_{n+k})\,(1+\beta)^j.$$

In particular,

$$\mathfrak{G}_{\sigma_k}^{(n)}(x_1 = 1, \ldots, x_{n+k} = 1) = \sum_{j=0}^{k-1} \binom{n+j-1}{j} \beta^j.$$

Problems 4.2. Give a bijective proof of Theorem 4.3, i.e., construct bijection between the set of k-tuple of non-crossing Schröder paths $(\mathfrak{p}_1, \ldots, \mathfrak{p}_k)$ of lengths $(n, n-1, \ldots, n-k+1)$ correspondingly <u>and</u> the set of pairs $(\mathfrak{m}, \mathcal{T})$, where $\mathcal{T}$ is a k-dissection of a convex $(n+k+1)$-gon, and $\mathfrak{m}$ is a upper triangle $(0,1)$-matrix of size $(k-1) \times (k-1)$, which is compatible with natural statistics on both sets.

4.3. *The "longest element" and Chan–Robbins polytope*

Assume additionally, cf. Ref. 28, **6.C8**, (d), that the condition (a) in Definition 4.1 is replaced by that:

(a') : x_{ij} and x_{kl} *commute* for all i, j, k and l.

Consider the element $w_0 := \prod_{1 \leq i < j \leq n} x_{ij}$. Let us bring the element w_0 to the reduced form, that is, let us consecutively apply the defining relations (a') and (b) to the element w_0 in any order until unable to do so. Denote the resulting polynomial by $Q_n(x_{ij}; \beta)$. Note that the polynomial itself <u>depends</u> on the order in which the relations (a') and (b) are applied.

We denote by $Q_n(\beta)$ the specialization $x_{ij} = 1$ for all i and j, of the polynomial $Q_n(x_{ij}; \beta)$, and by $Q_n(\beta, t)$ the specialization $x_{ij} = 1$, if $(i, j) \neq (1, n)$, and $x_{1,n} = t$,

Example 4.4. $Q_3(\beta) = (2, 1) = 1 + (\beta + 1)$, $Q_4(\beta) = (10, 13, 4) = 1 + 5(\beta + 1) + 4(\beta + 1)^2$, $Q_4(\beta, t) = t^4 + t (1 + 2t^2 + 2t^3) (\beta + 1) + (t + t^2)^2(\beta + 1)^2$, $Q_5(\beta) = (140, 336, 280, 92, 9) = 1 + 16(\beta + 1) + 58(\beta + 1)^2 + 56(\beta + 1)^3 + 9(\beta + 1)^4$, $Q_6(\beta) = 1 + 42(\beta + 1) + 448(\beta + 1)^2 + 1674(\beta + 1)^3 + 2364(\beta + 1)^4 + 1182(\beta + 1)^5 + 169(\beta + 1)^6$.

What one can say about the polynomial $Q_n(\beta) := Q_n(x_{ij}; \beta)|_{x_{ij}=1, \forall i,j}$?

It is known, in Ref. 28, **6.C8**, (d), that the constant term of the polynomial $Q_n(\beta)$ is equal to the product of Catalan numbers $\prod_{j=1}^{n-1} C_j$. It is not difficult to see that if $n \geq 2$, then $\mathrm{Coeff}_{[\beta+1]}(Q_n(\beta)) = 2^n - 1 - \binom{n+1}{2}$.

Theorem 4.4. *One has:*

$$Q_n(\beta - 1) = \left(\sum_{m \geq 0} \iota(CR_{n+1}, m)\beta^m \right) (1 - \beta)^{\binom{n+2}{2}+1},$$

where CR_m *denotes the Chan–Robbins polytope,*[2] *i.e., the convex polytope given by the following conditions:*

$CR_m = \{(a_{ij}) \in \mathrm{Mat}_{m \times m}(\mathbb{Z}_{\geq 0})\}$ *such that:*

(1) $\sum_i a_{ij} = 1, \sum_j a_{ij} = 1$,

(2) $a_{ij} = 0$, *if* $j > i + 1$.

Here for any integral convex polytope $\mathcal{P} \subset \mathbb{Z}^d$, $\iota(\mathcal{P}, n)$ *denotes the number of integer points in the set* $n\mathcal{P} \cap \mathbb{Z}^d$.

Conjecture 4.1.

(A) Let $n \geq 4$ and write

$$Q_n(\beta, t) := \sum_{k=0}^{2n-6} (1 + \beta)^k c_{k,n}(t), \quad then \ c_{k,n}(t) \in \mathbb{Z}_{\geq 0}[t].$$

(B) All roots of the polynomial $Q_n(\beta)$ belong to the set $\mathbb{R}_{<0}$.

Problems 4.3.

(1) Assume additionally to the conditions (a') and (b) above that:

$$x_{ij}^2 = \beta x_{ij} + 1, \quad if \ 1 \leq i < j \leq n.$$

What one can say about a reduced form of the element w_0 in this case?

(2) According to a result by Matsumoto and Novak,[22] if $\pi \in \mathbb{S}_n$ is a permutation of the cyclic type $\lambda \vdash n$, <u>then</u> the total number of <u>primitive factorizations</u> (see

definition in Ref. 22) of π into product of $n - \ell(\lambda)$ transpositions, denoted by $\mathrm{Prim}_{n-\ell(\lambda)}(\lambda)$, is equal to the product of Catalan numbers:

$$\mathrm{Prim}_{n-\ell(\lambda)}(\lambda) = \prod_{i=1}^{\ell(\lambda)} \mathrm{Cat}_{\lambda_i - 1}.$$

Recall that the Catalan number $\mathrm{Cat}_n := C_n = (1/n)\binom{2n}{n}$. Now take $\lambda = (2, 3, \ldots, n + 1)$. Then,

$$Q_n(1) = \prod_{a=1}^{n} \mathrm{Cat}_a = \mathrm{Prim}_{\binom{n}{2}}(\lambda).$$

Does there exist "a natural" bijection between the primitive factorizations and monomials which appear in the polynomial $Q_n(x_{ij}; \beta)$.

Acknowledgments

I thank Professor Toshiaki Maeno for many years fruitful collaboration. I am also grateful to Professors Y. Bazlov, I. Burban, B. Feigin, S. Fomin, A. Isaev, M. Ishikawa, B. Shapiro and Dr. Evgeny Smirnov for fruitful discussions on different stages of writing.[14]

References

1. Y. Bazlov, *J. Algebra* **297**(2), 372 (2006).
2. C. S. Chan and D. Robbins, *Exp. Math.* **8**(3), 291 (1999).
3. R. Brualdi and S. Kirkland, *J. Comb. Theory Ser. B* **9**(2), (2005).
4. C. Dunkl, *Trans. AMS* **311**, 167 (1989).
5. C. Dunkl, *Geometriae Dedicata* **32**, 157 (1989).
6. S. Fomin, S. Gelfand and A. Postnikov, *J. Am. Math. Soc.* **10**, 565 (1997).
7. S. Fomin and A. N. Kirillov, Quadratic algebras, Dunkl elements and Schubert calculus, in *Advances in Geometry*, eds. J.-L. Brylinski *et al.*, Prog. in Math. Vol. 172 (Birkhäuser, Boston, 1995), pp. 147–182.
8. S. Fomin and A. N. Kirillov, *Yang–Baxter Equation, Symmetric Functions and Grothendieck Polynomials*, preprint arXiv:hep-th/9306005.
9. S. Fomin and A. N. Kirillov, *Discrete Math.* **153**, 123 (1996).
10. S. Fomin and A. N. Kirillov, *J. Algebraic Comb.* **6**, 311 (1997).
11. V. Ginzburg, M. Kapranov and E. Vasserot, *Elliptic Algebras and Equivariant Elliptic Cohomology*, preprint arXiv:math/0001005.
12. K. Hikami and M. Wadati, *J. Math. Phys.* **44**(8), 3569 (2003).
13. Y. Kajihara and M. Noumi, *Indag. Math. (N.S.)* **14**(3–4), 395 (2003).
14. A. N. Kirillov, *On Some Quadratic Algebras, II*, preprint.
15. A. N. Kirillov and T. Maeno, *Europ. J. Comb.* **25**, 1301 (2004).
16. A. N. Kirillov and T. Maeno, *Int. Math. Res. Not. IMRN* **14**, 23 (2008).
17. A. N. Kirillov and T. Maeno, *A Note on Quantum K-Theory of Flag Varieties*, preprint.
18. A. N. Kirillov and T. Maeno, *Algebra i Analiz* **22**(3), 155 (2010); translation in *St. Petersburg Math. J.* **22**(3), 447 (2011).
19. A. Knutson, E. Miller and A. Yong, *J. Reine Angew. Math.* **630**, 1 (2009).

20. A. Lascoux and M.-P. Schützenberger, *Symmetry and Flag Manifolds, Invariant Theory*, Vol. 996 (Springer, LN., 1983), pp. 118–144.

21. I. Macdonald, *Notes on Schubert Polynomials*, Vol. 6, Publications du LACIM (Université du Québec à Montréal, 1991).

22. S. Matsumoto and J. Novak, *Primitive Factorizations, Jucys–Murphy Elements and Matrix Models*, preprint arXiv:1005.0151.

23. K. Meszaros, *Root Polytopes, Triangulations and the Subdivision Algebra, I*, preprint arXiv:0904.2194.

24. E. Mukhin, V. Tarasov and A. Varchenko, *Bethe Subalgebras of the Group Algebra of the Symmetric Group*, preprint arXiv:1004.4248.

25. A. Postnikov, On a quantum version of Pieri's formula, *Advances in Geometry*, eds. J.-L. Brylinski *et al.*, Progress in Math., Vol. 172 (Birkhäuser, Boston, 1995), pp. 371–383.

26. L. Serrano and C. Stump, *Generalized Triangulations, Pipe Dreams and Simplicial Spheres, FPSAC 2011* (DMTCS, 2011), pp. 885–896.

27. N. J. A. Sloan, *The on-line encyclopedia of integer sequences*, (2004), http://www.research.att.com/ njas/sequences/.

28. R. Stanley, *Catalan Addendum*, preprint, Version of 30 April 2011.

29. R. Stanley, *Enumerative Combinatorics*, Vol. 2 (Cambridge University Press, UK, 1999).

30. C. Stump, *J. Comb. Theory Ser. A* **118**(6), 1794 (2011).

31. R. Sulanke, *Electron. J. Comb.* **7**, 9 (2000).

32. M. Wachs, *J. Comb. Theory Ser. A* **40**, 276 (1985).

33. A. Woo, *Catalan numbers and Schubert polynomials for* $w = 1(n + 1)\ldots, 2$, preprint, arXiv:math/0407160.

Chapter 13

FINITE PROJECTIVE SPACES, GEOMETRIC SPREADS OF LINES AND MULTI-QUBITS

METOD SANIGA

Astronomical Institute, Slovak Academy of Sciences

SK-05960 Tatranská Lomnica, Slovak Republic

Given a $(2N-1)$-dimensional projective space over GF(2), PG$(2N-1,2)$, and its geometric spread of lines, there exists a remarkable mapping of this space onto PG$(N-1,4)$ where the lines of the spread correspond to the points and subspaces spanned by pairs of lines to the lines of PG$(N-1,4)$. Under such mapping, a nondegenerate quadric surface of the former space has for its image a nonsingular Hermitian variety in the latter space, this quadric being *hyperbolic* or *elliptic* in dependence on N being *even* or *odd*, respectively. We employ this property to show that generalized Pauli groups of N-qubits also form two distinct families according to the parity of N and to put the role of symmetric Pauli operators into a new perspective. The $N=4$ case is taken to illustrate the issue, due to its link with the so-called black-hole/qubit correspondence.

Keywords: Finite projective spaces; spreads of lines; Pauli groups of N-qubits.

Multiple qubit states play a key role in various fields of quantum information theory like quantum computing, coding and quantum error-correction (see, e.g., Ref. 1). Recently, and rather surprisingly, they have also been recognized to be of great relevance for getting insights into the nature of entropy formulas of a certain class of stringy black hole solutions (see, e.g., Ref. 2). It is, therefore, important to deepen our understanding of these fundamental buildings blocks of quantum world. In the present note we do so through the geometry of their associated generalized Pauli groups.

Let PG(d,q) be a d-dimensional projective space over GF(q), q being a power of a prime.[a] A t-spread $\mathcal{S}$ of PG(d,q) is a set of t-dimensional subspaces of PG(d,q) which partitions its point-set.[4] If the elements of $\mathcal{S}$ in a subspace V form a t-spread on V, one says that $\mathcal{S}$ induces a t-spread on V. A t-spread $\mathcal{S}$ is called *geometric* (or *normal*) if it induces a t-spread on each $(2t+1)$-dimensional subspaces

---

[a]For the standard mathematical nomenclature and notation employed in what follows, see, e.g., Ref. 3.

of PG(d, q) spanned by a pair of its elements. It is a well-known fact that PG(d, q) possesses a t-spread iff $(t + 1)|(d + 1)$; moreover, this condition is also sufficient for PG(d, q) to have a geometric t-spread. Segre showed[5] that a geometric t-spread of PG$(N(t + 1) - 1, q)$, $N \geq 2$, gives rise to a projective space PG$(N - 1, q^{t+1})$ as follows: the points of this space are the elements of $\mathcal{S}$ and its lines are the $(2t + 1)$-dimensional subspaces spanned by any two distinct elements of $\mathcal{S}$, with incidence inherited from PG$(N(t + 1) - 1, q)$. For a particular case of $t = 1$ (i.e., a spread of lines), Dye[6] demonstrated that a *hyperbolic* or an *elliptic* quadric of PG$(2N - 1, q)$ has an induced (geometric) spread of lines if and only if N is, respectively, *even* or *odd*, in which case it is mapped onto a nonsingular Hermitian variety H$(N-1, q^2)$ of PG$(N-1, q^2)$. We shall now show that this property has for $q = 2$ a very interesting physical implication.

It is already a firmly established fact[7–10] that the commutation relations between the elements of the generalized Pauli group of N-qubits, $N \geq 2$, can be completely reformulated in the geometrical language of symplectic polar space of rank N and order two, W$(2N - 1, 2)$; the generalized Pauli operators (discarding the identity) answer to the points of W$(2N - 1, 2)$, a maximally commuting subset has its representative in a maximal totally isotropic subspace of W$(2N - 1, 2)$ and commuting translates into collinear. One of the most natural representations of W$(2N-1, 2)$ is that in terms of the points and the set of totally isotropic subspaces of PG$(2N-1, 2)$ endowed with a symplectic polarity. Employing this representation, it has been found in Ref. 8 that in the real case the *symmetric* elements/operators of the N-qubit Pauli group *always* lie on a *hyperbolic* quadric in the ambient space PG$(2N - 1, 2)$. Combining this fact with Dye's result, we arrive at our main observation: *it is only for N even when all symmetric generalized Pauli operators of W$(2N - 1, 2)$ can be mapped to the points of an Hermitian variety of the space PG$(N - 1, 4)$ associated through a geometric spread of lines with the ambient space PG$(2N - 1, 2)$*. Hence, in this regard, when it comes to generalized Pauli groups "even-numbered" multi-qubits are found to stand on a slightly different footing than "odd-numbered" ones.

We shall finish this paper by briefly mentioning an especially interesting even case, $N = 4$. Here, a hyperbolic quadric Q$^+(7, 2)$ of PG$(7, 2)$ formed by the symmetric operators is well-known for its puzzling triality swapping points and two systems of generators and has for its spread-induced image an Hermitian surface H$(3, 4)$ of PG$(3, 4)$ (see, e.g., Ref. 11). This Hermitian surface is, in turn, nothing but the generalized quadrangle GQ$(4, 2)$ in disguise (see, e.g., Refs. 12 and 13), the dual of which — GQ$(2, 4)$ — was found to play a prominent role in the so-called black-hole-qubit correspondence, by fully encoding the $E_{6(6)}$ symmetric entropy formula describing black holes and black strings in $D = 5$.[14] Our finding thus, *inter alia*, not only opens up an unexpected window through which also four-qubit Pauli group, like its lower rank cousins, could find its way into some black hole entropy formula(s), but also puts the role of symmetric operators into a new perspective.

It is also important to keep in mind this remarkable *three-to-one* correspondence, i.e., that it is always a triple of (collinear) operators of the ambient space $PG(7,2)$ which comprises a single point of $PG(3,4)$.

Acknowledgments

This work was partially supported by the VEGA grant agency, projects Nos. 2/0092/09 and 2/0098/10. The idea exposed in this paper originated from discussions with Prof. Hans Havlicek, Dr. Boris Odehnal (Vienna University of Technology) and Dr. Petr Pracna (J. Heyrovský Institute of Physical Chemistry, Prague).

References

1. M. A. Nielsen and I. L. Chuang, *Quantum Computation and Quantum Information* (Cambridge University Press, Cambridge, 2000).
2. L. Borsten *et al.*, *Phys. Rep.* **471**, 113 (2009) arXiv:0809.4685.
3. J. W. P. Hirschfeld and J. A. Thas, *General Galois Geometries* (Oxford University Press, Oxford, 1991).
4. J. Eisfeld and M. Storme, (Partial) *t*-spreads and minimal *t*-covers in finite projective spaces. Lecture notes available from http://cage.rug.ac.be/˜fdc/intensivecourse2/storme2.ps.
5. B. Segre, *Ann. Mat. Pura Appl.* **64**, 1 (1964).
6. R. H. Dye, *J. Lond. Math. Soc.* **33**, 279 (1986).
7. M. Saniga and M. Planat, *Adv. Studies Theor. Phys.* **1**, 1 (2007) arXiv:quant-ph/0612179.
8. H. Havlicek, B. Odehnal and M. Saniga, *SIGMA* **5**, 096 (2009) arXiv:0903.5418.
9. K. Thas, *Europhys. Lett.* **86**, 60005 (2009).
10. P. Vrana and P. Lévay, *J. Phys. A: Math. Theor.* **43**, 125303 (2010) arXiv:0906.3655.
11. H. Havlicek, B. Odehnal and M. Saniga, *Des. Codes Cryptogr.* **62**, 343 (2012) arXiv:1006.4492.
12. S. E. Payne and J. A. Thas, *Finite Generalized Quadrangles* (Pitman, Boston, London, Melbourne, 1984).
13. K. Thas, *Symmetry in Finite Generalized Quadrangles* (Birkhäuser, Basel, 2004).
14. P. Lévay *et al.*, *Phys. Rev. D* **79**, 084036 (2009) arXiv:0903.0541.

Chapter 14

MONOGAMY OF ENTANGLEMENT, N-REPRESENTABILITY PROBLEMS AND GROUND STATES

TZU-CHIEH WEI

C. N. Yang Institute for Theoretical Physics,
State University of New York at Stony Brook,

Stony Brook, NY 11794-3840, USA

Monogamy of entanglement is a quantum mechanical property that limits quantum correlations shared among many parties. In an example strongly interacting spin system
we examine approaches for approximating the ground state energy both from above and
below by mean-field and N-representability methods, respectively. Due to strong competition among the terms in the Hamiltonian, the resulting ground-state wavefunction,
although is entangled, does not possess entanglement that is proportional to the system
size, thus obeying the monogamy of entanglement.

Keywords: Monogamy of entanglement; N-representability; ground-state energy; spin
systems.

1. Introduction

Study of quantum spin chains dates back to the early twentieth century. The famous
examples include Bethe's solution for Heinsenberg spin chains.[1] In recent decades,
quantum information emerges as a new branch of science,[2,3] and quantum spins
serve as important elements in both theoretical and experimental quantum information sciences. Solving ground-state energy for both class and quantum spin systems can be a challenge.[3,4] Standard techniques involve variational methods, which
approach the ground state energy from above. In quantum chemistry, the so-called
N-representability problem actually approach the ground-state energy from below,
complementing the variational approach. Here we give an example Hamiltonian,
which is an example of quantum satisfiability problem (QSAT),[5,6] and illustrate
the above two aspects in the search of the ground state and its energy.[7]

Consider Hamiltonian of n qubits (spin-1/2 particles) residing on the vertices
of a complete graph C_n; see Fig. 1:

$$H = -\sum_{\langle i,j \rangle} |\psi^+\rangle \langle \psi^+|_{ij} , \qquad (1)$$

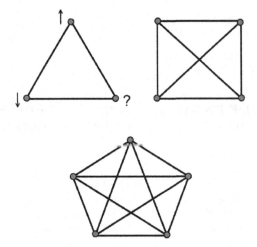

Fig. 1.   (Examples of complete graphs.) The spins reside on the vertices, indicated by solid circles. The edges of the graph represent interaction between the corresponding two spins at the ends. In general, the Hamiltonian is frustrated, as the term $|\psi^+\rangle\langle\psi^+|$ tries to force the two spins correlated (or more precisely, entangled) in opposite directions. We shall also see that $|\psi^-\rangle\langle\psi^-|$ interaction is more frustrated.

where $\langle i, j \rangle$ denotes the edge that connects vertices i and j, $|\psi^+\rangle_{ij} = (|01\rangle_{ij} + |10\rangle_{ij})/\sqrt{2}$, with $|0\rangle \equiv |\uparrow\rangle$ and $|1\rangle \equiv |\downarrow\rangle$. Each term $-|\psi^+\rangle\langle\psi^+|_{ij}$ tries to force the corresponding two qubits (i and j) to be in the state $|\psi^+\rangle_{ij}$. However, if i and j form the triplet state, the maximal entanglement of two qubits, they each cannot have entanglement with other qubits. Such phenomenon is called the *monogamy of entanglement*.[8,9] The Hamiltonian is highly frustrated.

We note that $-|\psi^+\rangle\langle\psi^+|_{ij}$ can be expressed in terms of Pauli matrices of spins i and j as follows,

$$- |\psi^+\rangle\langle\psi^+|_{ij} = -\frac{1}{4}\left(\sigma_x^{[i]}\sigma_x^{[j]} + \sigma_y^{[i]}\sigma_y^{[j]} - \sigma_z^{[i]}\sigma_z^{[j]} + I\right). \tag{2}$$

For nearest-neighbor interactions of such form on 1D chain, it is equivalent to the antiferromagnetic Heisenberg chain. Furthermore, adding an identity to each term make it an projection operator: $I - |\psi^+\rangle\langle\psi^+|$, and the problem is identified as a quantum 2-SAT problem.[5]

2. Mean-Field Approximation

The Hamiltonian (1) is invariant under permutation and, in particular, $|\psi^+\rangle_{ij} = |\psi^+\rangle_{ji}$. We can use the mean-field state $|\phi^{\otimes n}\rangle$ to derive an upper bound on the ground-state energy:

$$E_0 \leq \min\langle\phi^{\otimes n}|H|\phi^{\otimes n}\rangle = -n_{\text{pair}}|\langle\phi\phi|\psi^+\rangle|^2, \tag{3}$$

where $n_{\text{pair}} = n(n-1)/2$. The minimum is attained, e.g., when $|\phi\rangle = |+\rangle$, and we arrive at the upper bound on the ground-state energy

$$E_0 \leq -n_{\text{pair}}/2. \tag{4}$$

We shall see below that the upper bound is actually a very good approximation to the ground-state energy, as expected for highly connected interaction graph. However, the product wavefunction $|+^{\otimes n}\rangle$ has a poor resemblance to the true ground-state wavefunction.

3. *N*-Representability Problem

Roughly speaking, given a set of generally mixed states $\{\rho^{[i_1,i_2,\dots,i_k]}\}$, the N-representability problem[10-14] inquires whether they come from the same N-particle state by tracing over the remaining $N-k$ particles (where N is the total number of particles in the system), i.e.,

$$\rho^{[i_1,i_2,\dots,i_k]} = \text{Tr}_{N-k}(\rho), \tag{5}$$

where ρ is an N-particle state, which can be pure or mixed. It turns out that N-representability is a QMA-hard problem for spins,[15] fermions[16] and bosons.[17] In particular, the solutions of N-representability problems give rise to solutions of many-body ground-state energy.

Let us consider a Hamiltonian consisting of two-body interacting terms h_{ij}, i.e., $H = \sum_{[i,j]} h_{ij}$, where the sum is over certain set of pairs $[i,j]$, depending on the geometry, graph or physical settings at hand. If one can, given $\{\rho^{[i,j]}\}$, determine whether there exists ρ_N such that $\rho^{[i,j]} = \text{Tr}_{N-[i,j]}\rho_N$, where ρ_N is an N-particle state, then the ground-state energy $E_0 = \min_\psi \langle\psi|H|\psi\rangle$ can be solved,

$$E_0 = \min_\psi \langle\psi|H|\psi\rangle = \min_\psi \sum_{[i,j]} \text{Tr}(|\psi\rangle\langle\psi|h_{ij}) \tag{6}$$

$$= \min_{\rho^{[i,j]} \, N\text{-representable}} \text{Tr}\left(\rho^{[i,j]} h_{ij}\right). \tag{7}$$

If we do not impose the N-representability of the reduced two-particle states $\rho^{[i,j]}$'s or if we relax the constraint, the right-hand side will give a *lower* bound on the ground state energy.

Because of the particular form of the Hamiltonian (1) considered, we expect that the ground state to possess the permutation invariance as the Hamiltonian itself. In the following, we shall see how one can apply N-representability to our problem.

3.1. *3-representable*

Without any constraint on the reduced two-spin states in Eq. (7), we easily obtain a lower bound on the ground-state energy, i.e., $E_0 \geq -n_{\text{pair}}$. By using the result of

Brayvi[5] we know that this lower bound is not the ground-state energy, as there is no state that can satisfy all the two-body terms (determining the existence or not is a problem in P). Including constraints, we can hence push upward the value of the lower bound, but it may also become harder to solve.

We are going to consider only permutation symmetric states. For n qubits, these can be classified by the basis states $|S(n,k)\rangle$'s in the symmetric subspace:

$$|S(n,k)\rangle = \frac{\sqrt{C_k^n}}{n!} \sum_{\Pi_i \in S_n} |\Pi_i(\underbrace{0,\ldots,0}_{n-k}, \underbrace{1,\ldots,1}_{k})\rangle, \tag{8}$$

where $C_k^n \equiv n!/(n!(n-k)!)$, and Π_i is an group element in the symmetric group S_n. Any permutation invariant pure quantum state $|\psi\rangle$ can be written as:

$$|\psi_n\rangle = \sum_{k=0}^{n} a_k |S(n,k)\rangle. \tag{9}$$

For example, in the case $n = 3$, we have

$$|S(3,0)\rangle = |000\rangle, |S(3,3)\rangle = |111\rangle, \tag{10}$$

$$|S(3,1)\rangle = (|001\rangle + |010\rangle + |100\rangle)/\sqrt{3}, \tag{11}$$

$$|S(3,2)\rangle = (|110\rangle + |101\rangle + |011\rangle)/\sqrt{3}. \tag{12}$$

Suppose we consider $|\psi_3\rangle \equiv \sum_{k=0}^{3} a_k |S(3,k)\rangle$ and trace it over one spin, we obtain:

$$\rho_2 = \text{Tr}_1 |\psi_3\rangle\langle\psi_3| = |\alpha\rangle\langle\alpha| + |\beta\rangle\langle\beta|, \tag{13}$$

$$|\alpha\rangle \equiv a_0|00\rangle + \frac{a_2}{\sqrt{3}}|11\rangle + \frac{a_1}{\sqrt{3}}(|01\rangle + |10\rangle), \tag{14}$$

$$|\beta\rangle \equiv a_3|11\rangle + \frac{a_1}{\sqrt{3}}|00\rangle + \frac{a_2}{\sqrt{3}}(|01\rangle + |10\rangle), \tag{15}$$

noting that $|\alpha\rangle$ and $|\beta\rangle$ are not normalized to unity. The above two-spin state ρ_2 is derived from tracing one spin over a three-spin state and hence it is 3-representable. Using this two-spin (3-representable) state and minimizing the energy over a's, we obtain a lower bound on the ground-state energy,

$$E_0 \geq \min_{a's} n_{\text{pair}}(-1)\text{Tr}(\rho_2|\psi^+\rangle\langle\psi^+|) \tag{16}$$

$$\geq \min_{a's} n_{\text{pair}}(-1)\frac{2}{3}(|a_1|^2 + |a_2|^2) = -\frac{2}{3}n_{\text{pair}}. \tag{17}$$

The lower bound is now pushed from $-n_{\text{pair}}$ to $-2n_{\text{pair}}/3$.

One can continue by using four-representable states or higher, and push the lower bound even further up.

3.2. *n-representable*

We can actually employ the full n-representable states by considering $\rho_2' \equiv \text{Tr}_{n-2}|\psi_n\rangle\langle\psi_n|$, where $|\psi_n\rangle$ is a general n-qubit symmetric state (9). By using the identity

$$|S(n,k)\rangle = |00\rangle\sqrt{\frac{C_k^{n-2}}{C_k^n}}|S(n-2,k)\rangle + (|01\rangle+|10\rangle)\sqrt{\frac{C_{k-1}^{n-2}}{C_k^n}}|S(n-2,k-1)\rangle$$

$$+ |11\rangle\sqrt{\frac{C_{k-2}^{n-2}}{C_k^n}}|S(n-2,k-2)\rangle, \tag{18}$$

we can calculate the reduced two-spin state ρ_2' and derive the expression for the exact ground-state energy

$$E_0 = \min_{a's} n_{\text{pair}}(-1)\text{Tr}(\rho_2'|\psi^+\rangle\langle\psi^+|) \tag{19}$$

$$= \min_{a's} n_{\text{pair}}(-1)\sum_k |a_k|^2 \frac{2k(n-k)}{n(n-1)}. \tag{20}$$

For n even, the minimum is achieved at $k = n/2$,

$$E_0/n_{\text{pair}} = -\frac{1}{2}\frac{n}{n-1}, \tag{21}$$

and the ground state is $|S(n,n/2)\rangle$. There is an finite energy gap to the ground state (in the thermodynamic limit). For n being odd, the minimum is achieved at $k = (n-1)/2$ or $k = (n+1)/2$,

$$E_0/n_{\text{pair}} = -\frac{1}{2}\frac{n}{n-1}\left(1 - \frac{1}{n^2}\right), \tag{22}$$

and there is also an energy gap above the degenerate ground space.

3.3. *Ground-state entanglement*

From the mean-field approximation, we obtain $|+^{\otimes n}\rangle$ for minimizing the energy. For large n, the overlap of it with the ground state,

$$|\langle +^{\otimes n}|E_0\rangle|^2 \sim \sqrt{\frac{2}{\pi n}}, \tag{23}$$

becomes vanishing small. On the other hand, entanglement of permutation invariant states have also been investigated before.[18–22] In terms of the geometric entanglement,[18] the ground-state entanglement (for even and large n) is $(1/2)\log(\pi n/2)$. The entanglement per spin for large n is vanishing small, i.e., $\sim \log(n)/n$.

4. Concluding Remarks

We studied an example strongly interacting spin system and examined approaches for approximating the ground-state energy both from above and below by mean-field and N-representability methods, respectively. Due to the monogamy of entanglement, the ground state cannot be maximally entangled states, i.e., does not achieve the maximal entanglement per spin. Indeed, we have seen that the ground state possess the amount of entanglement logarithmical in the total number of spins. The Hamiltonian we consider, when placed on arbitrary graph, is not generally solvable.

However, for the complete graphs we considered, it is exactly solvable. The Hamiltonian (1) can be rewritten,

$$H = \sum_{\langle i,j \rangle} -\frac{1}{4}\left(\sigma_x^{[i]}\sigma_x^{[j]} + \sigma_y^{[i]}\sigma_y^{[j]} - \sigma_z^{[i]}\sigma_z^{[j]} + I\right) \tag{24}$$

$$= \frac{2n - n^2}{8} - \frac{1}{2}\left(S_X^2 + S_Y^2 - S_Z^2\right), \tag{25}$$

$$= \frac{2n - n^2}{8} - \frac{1}{2}\left(\hat{S}^2 - 2S_Z^2\right), \tag{26}$$

where $S_X \equiv \sum_i \sigma_x^{[i]}/2$, $S_Y \equiv \sum_i \sigma_y^{[i]}/2$, $S_Z \equiv \sum_i \sigma_z^{[i]}/2$ and $\hat{S} \equiv (S_X, S_Y, S_Z)$.

To minimize the energy, we need S_Z as small and $\hat{S}^2$ as large as possible. For n even, this is achieved for $S = n/2$, hence $\hat{S}^2 = (n/2)((n/2)+1)$ and $S_z = 0$. The ground state is thus characterized by the total spin S and total z-component spin S_z as $|S = n/2, S_z = 0\rangle$, the same as $|S(n, n/2)\rangle$. For n odd, this is achieved for $S = n/2$ and $S_z = \pm 1/2$ and the ground state is doubly degenerate $|S = n/2, S_z = \pm 1/2\rangle = |S(n, (n \pm 1)/2)\rangle$.

In fact, the above approach can be used to deal with other types of two-body terms, such as $|\phi^\pm\rangle\langle\phi^\pm|$ and $|\psi^-\rangle\langle\psi^-|$, where $|\phi^\pm\rangle \equiv (|00\rangle \pm |11\rangle)/\sqrt{2}$ and $|\psi^-\rangle \equiv (|01\rangle - |10\rangle)$. The former two cases will give results similar to the above, whereas the latter ($|\psi^-\rangle$) gives rise to a different outcome. This is because the state $|\psi^-\rangle$ is antisymmetric in exchanging the two spins and thus such a term $|\psi^-\rangle\langle\psi^-|$ in the Hamiltonian will work to enforce the antisymmetry.

In this case, the Hamiltonian is:

$$H = \sum_{\langle i,j \rangle} \frac{1}{4}\left(\sigma_x^{[i]}\sigma_x^{[j]} + \sigma_y^{[i]}\sigma_y^{[j]} + \sigma_z^{[i]}\sigma_z^{[j]} - I\right) \tag{27}$$

$$= \frac{-2n - n^2}{8} + \frac{1}{2}\hat{S}^2. \tag{28}$$

For n even, the ground state is $|S = 0, S_z = 0\rangle$ and has energy $E_0 = -(2n + n^2)/8$. For n odd, the ground state is degenerate $|S = 1/2, S_z = \pm 1/2\rangle$ and has energy $E_0 = (3 - 2n - n^2)/8$. The ground-state energy is apparently higher than the case of triplet interactions, and this means the singlet interaction is more frustrated. For one-dimensional chains, however, the four different interactions (three triplet and one singlet) give rise to the same ground-state energy.

Acknowledgment

The author acknowledges support from the C. N. Yang Institute for Theoretical Physics. It is a great pleasure for the author to dedicate this paper to his colleague, Vladimir Korepin, as a contribution for the book celebrating the occasion of his 60th birthday.

References

1. V. E. Korepin, N. M. Bogoliubov and A. G. Izergin, *Quantum Inverse Scattering Method and Correlation Functions* (Cambridge University Press, Cambridge, 1993).
2. M. Nielsen and I. Chuang, *Quantum Computation and Quantum Information* (Cambridge University Press, Cambridge, 2000).
3. A. Yu. Kitaev, A. H. Shen and M. N. Vyalyi, *Classical and Quantum Computation* (AMS, Providence, 2002).
4. F. Barahona, *J. Phys. A: Math. Gen.* **15**, 3241 (1982).
5. S. Bravyi, e-print quant-ph/0602108.
6. C. R. Laumann *et al.*, *Quantum Inf. Comput.* **10**, 1 (2010).
7. F. Verstraete and J. I. Cirac, *Phys. Rev. B* **73**, 094423 (2006).
8. V. Coffman, J. Kundu and W. K. Wootters, *Phys. Rev. A* **61**, 052306 (2000).
9. T. J. Osborne and F. Verstraete, *Phys. Rev. Lett.* **96**, 220503 (2006).
10. A. J. Coleman, *Rev. Mod. Phys.* **35**, 668 (1963).
11. C. A. Coulson, *Rev. Mod. Phys.* **32**, 175 (1960).
12. R. H. Trethold, *Phys. Rev.* **105**, 1421 (1957).
13. A. J. Coleman and V. I. Yukalov, *Reduced Density Matrices: Coulson's Challenge* (Springer, Berlin, 2000).
14. J. Cioslowski, *Many-Electron Densities and Reduced Density Matrices* (Springer, Berlin, 2000).
15. Y.-K. Liu, The complexity of the consistency and N-representability problems for quantum states, Ph.D. Thesis, Unversity of California–San Diego (2007); also in e-print arXiv:0712.3041.
16. Y.-K. Liu, M. Christandl and F. Verstraete, *Phys. Rev. Lett.* **98**, 110503 (2007).
17. T.-C. Wei, M. Mosca and A. Nayak, *Phys. Rev. Lett.* **104**, 040501 (2010).
18. T.-C. Wei and P. M. Goldbart, *Phys. Rev. A* **68**, 042307 (2003).
19. T.-C. Wei, *Phys. Rev. A* **78**, 012327 (2008).
20. R. Hübener *et al.*, *Phys. Rev. A* **80**, 032324 (2009).
21. M. Hayashi *et al.*, *J. Math. Phys.* **50**, 122104 (2009).
22. T.-C. Wei and S. Severini, *J. Math. Phys.* **51**, 092203 (2010).

SCIENTIFIC PROGRAM

Pre-registration: 24 May 2011 (Tuesday)	
04:00 pm – 08:00 pm	Pre-Registration

Day 1 Program: 25 May 2011 (Wednesday)	
08:00 am	Registration
08:45 am – 09:00 am	Welcome address by **Prof Phua Kok Khoo (Director, IAS)** Opening address by **Prof Artur Ekert (Director, CQT)**

Morning Session I
Chairman: **Artur Ekert**

09:00 am – 09:30 am	**Franco Nori** (RIKEN and Univ. of Michigan) Photons interacting with solid state qubits
09:30 am – 10:00 am	**Zidan Wang** (Univ. of Hong Kong) Topological superconductors and Dirac fermions: exact results
10:00 am – 10:30 am	**Darrick Chang** (Caltech) Slowing and stopping light using an optomechanical crystal array
10:30 am – 11:00 am	**Group Photo and Tea Break**

Morning Session II
Chairman: **Franco Nori**

11:00 am – 11:30 am	**Wu-Ming Liu** (Institute of Physics, Beijing) Non-Abelian Josephson effect
11:30 am – 12:00 pm	**Jaewan Kim** (Korea Institute for Advanced Study) Optical qudit cluster state and teleportation of qudits
12:00 pm – 12:30 pm	**Xiwen Guan** (Australian National Univ.) Fermi polaron, pair correlation and scaling in 1D interacting fermions
12:30 pm – 02:00 pm	**Lunch Break**

Afternoon Session I
Chairman: **Bei-Lok Hu**

02:00 pm – 02:30 pm	**Howard Wiseman** (Griffith Univ.) The power of many settings or many outcomes in experimental demonstrations of EPR-steering
02:30 pm – 03:00 pm	**Fuli Li** (Xi'an Jiaotong Univ.) Controlling quantum entanglement and discord evolution of spin systems by multi-body interaction and external magnetic field
03:00 pm – 03:30 pm	**Wang Yao** (Univ. of Hong Kong) Quantum state engineering of spins by collective spin pumping
03:30 pm – 04:00 pm	**Yang Yu** (Nanjing Univ.) Landau-Zener-Stückelberg interference in a superconducting phase qubit
04:00 pm – 04:30 pm	**Tea Break**

Afternoon Session II
Chairman: **Howard Carmichael**

04:30 pm – 05:00 pm	**Wei-Min Zhang** (National Cheng Kung Univ.) Quantum transport theory for photonic networks
05:00 pm – 05:30 pm	**Jiangfeng Du** (Univ. of Sci. & Technol. of China) Preservation of bipartite pseudo-entanglement in solids using dynamical decoupling
05:30 pm – 06:00 pm	**Zhengfu Han** (Univ. of Sci. & Technol. of China) Differential phase shift quantum key distribution with continuous wave laser
06:00 pm – 06:30 pm	**Hsi-Sheng Goan** (National Taiwan Univ.) Optimal control of quantum gates for a non-Markovian open quantum bit system
06:30 pm – 08:00 pm	**Catered Dinner**

Day 2 Program: 26 May 2011 (Thursday)

Morning Session I
Chairman: **Wu-Ming Liu**

09:00 am – 09:30 am	**Bei-Lok Hu** (Univ. of Maryland) Non-Markovian entanglement dynamics of atom-field systems
09:30 am – 10:00 am	**Jianqiang You** (Fudan Univ.) A quantum entanglement approach for topological quantum phase transitions
10:00 am – 10:30 am	**Barry Sanders** (Univ. of Calgary) Empirically discerning Autler-Townes splitting from electromagnetically induced transparency
10:30 am – 11:00 am	**Tea Break**

Morning Session II
Chairman: **Zidan Wang**

11:00 am – 11:30 am	**Hai-Qing Lin** (Chinese Univ. of HK) The role of low-lying excitations on the ground state entanglement
11:30 am – 12:00 pm	**Weiping Zhang** (East China Normal Univ.) Dynamics of a Bose-Einstein condensate in a self-distorted optical lattice
12:00 pm – 12:30 pm	**Tiancai Zhang** (Shanxi Univ.) Elimination of degenerate trajectory of single atom strongly coupled to the high order cavity mode
12:30 pm – 02:00 pm	**Poster Session and Lunch Break**

Afternoon Session I
Chairman: **Barry Sanders**

02:00 pm – 02:30 pm	**Howard Carmichael** (Univ. of Auckland) Elastic light scattering in multi-level atoms: evolution of ground-state coherence and quantum measurement through quantum jumps
02:30 pm – 03:00 pm	**Ren-Bao Liu** (Chinese Univ. of HK) Control of electron spin decoherence in solids and applications
03:00 pm – 03:30 pm	**Sixia Yu** (CQT/NUS and Univ. of Sci. & Technol. of China) The quantum discord: its detection and bounds

03:30 pm – 04:00 pm	**Ping Koy Lam** (Australian National Univ.) Unconditional room temperature quantum memory
04:00 pm – 04:30 pm	**Tea Break**

Afternoon Session II
Chairman: **Hai-Qing Lin**

04:30 pm – 05:00 pm	**Yan Chen** (Fudan Univ.) Characterizing boson density wave and valence bond orders in a lattice by its dual vortex degree of freedoms
05:00 pm – 05:30 pm	**Zhanghai Chen** (Fudan Univ.) Selective Purcell effect of spin polarized exciton emissions in a quantum dot – microcavity system
05:30 pm – 05:45 pm	**N.D.Hari Dass** (Indian Institute of Science & Chennai Mathematical Institute) Statistical significance of single unknown harmonic oscillator coherent states
05:45 pm – 06:00 pm	**Kaige Wang** (Beijing Normal Univ.) Two-photon talbot self-imaging
07:30 pm – 10:00 pm	**Banquet (by invitation only)**

Day 3 Program: 27 May 2011 (Friday)

Morning Session I
Chairman: **Molin Ge**

09:00 am – 09:30 am	**Rodney James Baxter** (Australian National Univ.) The Ising and chiral Potts models
09:30 am – 10:00 am	**Vladimir Korepin** (State Univ. of New York) Measures of entanglement in spin chains
10:00 am – 10:30 am	**Kazuo Fujikawa** (Nihon Univ.) Separability criterion for continuous variable systems and an analytic solution
10:30 am – 11:00 am	**Tea Break**

Morning Session II
Chairman: **Kazuo Fujikawa**

11:00 am - 11:30 am	**Molin Ge** (Nankai Univ.) Yang Baxter equation and quantum information
11:30 am – 12:00 pm	**Thomas Durt** (Ecole Centrale de Marseille, Institut Fresnel) About optimal cloning and entanglement
12:00 pm – 12:30 pm	**Alexei Tsvelik** (Brookhaven National Laboratory) A possible realization of zero energy Majorana modes in spin ladders
12:30 pm – 2:00 pm	**Lunch Break**

Afternoon Session I
Chairman: **Jianqiang You**

02:00 pm – 02:30 pm	**Xiang-Bin Wang** (Tsinghua Univ.) Improving quantum entanglement through single-qubit operation
02:30 pm – 03:00 pm	**Thibault Vogt** (Peking Univ.) Manipulation of a quantum gas: state, momentum and phase
03:00 pm – 03:30 pm	**Shunlong Luo** (Academy of Mathematics and Systems Science) Measurement-induced nonlocality

03:30 pm – 04:00 pm	**Masahito Hayashi** (Tohoku Univ.) Weaker entanglement guarantees stronger entanglement
04:00 pm – 04:30 pm	**Tea Break**

Afternoon Session II
Chairman: **Leong Chuan Kwek**

04:30 pm – 05:00 pm	**Vladislav Popkov** (Univ. of Salerno) Entanglement of permutational invariant pure quantum states and mixtures
05:00 pm – 05:30pm	**Igor Volovich** (Steklov Mathematical Institute) Quantum photosynthesis and entropy decreasing
05:30 pm – 05:45 pm	**Keshav Narain Shrivastava** (Univ. of Malaya) Hubbard model, Peierls-Luttinger phase and Kohn's cyclotron resonance theorem
05:45 pm – 06:00 pm	**Peng Xue** (Southeast Univ.) Quantum Walks with Many Particles
06:00 pm – 08:00 pm	**Catered Dinner**

Day 4 Program: 28 May 2011 (Saturday)

Morning Session I
Chairman: **Jaewan Kim**

09:00 am – 09:30 am	**Paul Wiegmann** (Univ. of Chicago) Non-linear equation for Fermi gas
09:30 am – 10:00 am	**Hosho Katsura** (Gakushuin Univ.) Entanglement spectrum of valence-bond-solid states: VBS/CFT correspondence
10:00 am – 10:30 am	**Howard E. Brandt** (U.S. Army Research Laboratory) Aspects of the Riemannian geometry of quantum computation
10:30 am – 11:00 am	**Tea Break**

Morning Session II
Chairman: **Thomas Durt**

11:00 am – 11:30 am	**Luigi Amico** (Univ. of Catania) Ground state factorization and correlations in a quantum many body system
11:30 am – 12:00 pm	**Sougato Bose** (Univ. College London) Scattering for Quantum Information Processing
12:00 pm – 12:15 pm	**Daniel Valente** (Institut Néel CNRS France) Stimulated emission at the single photon level in 1D atoms
12:15 pm – 12:30 pm	**Syed M Assad** (Australian National University) Real time demonstration of high bitrate quantum random number generation with coherent laser light
12:30 pm – 01:30 pm	**Lunch Break**
02:30 pm – 09:30 pm	**City Tour**

End of Program

~ We wish all a safe journey home! ~

LIST OF PARTICIPANTS

Isaac Onoja Akogwu
University Tun Abdul Razak,
Malaysia

AN Junhong
CQT, National University Singapore,
Singapore

Muhammad Sabieh ANWAR
Lahore University of Management Sciences,
Pakistan

CHEN Qing
CQT, National University Singapore,
Singapore

DAI Li CQT, National University Singapore,
Singapore

Sahar HEJAZI
Shahid Beheshti University,
Iran

HU Yuxin
National University Singapore–CQT,
Singapore

Florian Hudelist
East China Normal University,
China

HUO Mingxia
National University Singapore–CQT,
Singapore

JIANG Nian-Quan
Wenzhou University,
China

KOH Chee Yeong (XU Zhiyong)
Nanyang Technological University Student,
Singapore

LEE Chee Kong
CQT–National University Singapore,
Singapore

LEE Kean Loon
National University Singapore–CQT,
Singapore

LI Hongwei
University of Science and Technology of China,
China

LI Ying
National University Singapore–CQT,
Singapore

Koji NAKATOGAWA
Hokkaido University,
Japan

Changsuk NOR
National University Singapore–CQT,
Singapore

ONG Wei Guang
Nanyang Technological University Student,
Singapore

Francis Norman Claridades PARAAN
State University of New York at Stony Brook,
United States

Andrii Petrashyk
Nanyang Technological University Staff – RES,
Singapore

Qian Jun
CQT–National University Singapore,
Singapore

Philippe RAYNAL
National University Singapore–CQT,
Singapore

SHANG Jiangwei
CQT, China

Wonmin Son
Sogang University,
Korea

SU Haibin
Nanyang Technological University-MSE/MS Div,
Singapore

TANG Chi Sin
Nanyang Technological University Student,
Singapore

TEO Yong Siah
National University Singapore–CQT,
Singapore

TIAN Feng Mei
CQT,
Singapore TONG Qingjun
National University Singapore,
Singapore

WANG Ming-Feng
Wenzhou University,
China

WEI Zhaohui
CQT, National University Singapore,
Singapore

WU Chunfeng
National University Singapore,
Singapore

YOU Jiabin
CQT–National University Singapore,
Singapore

ZHANG Yongsheng
University Science and Techology of China,
China

ZHANG Chengjie
Research Fellow of Professor Oh's Group-National University Singapore,
Singapore